Mercedes-Benz

CLASSIC LIFE

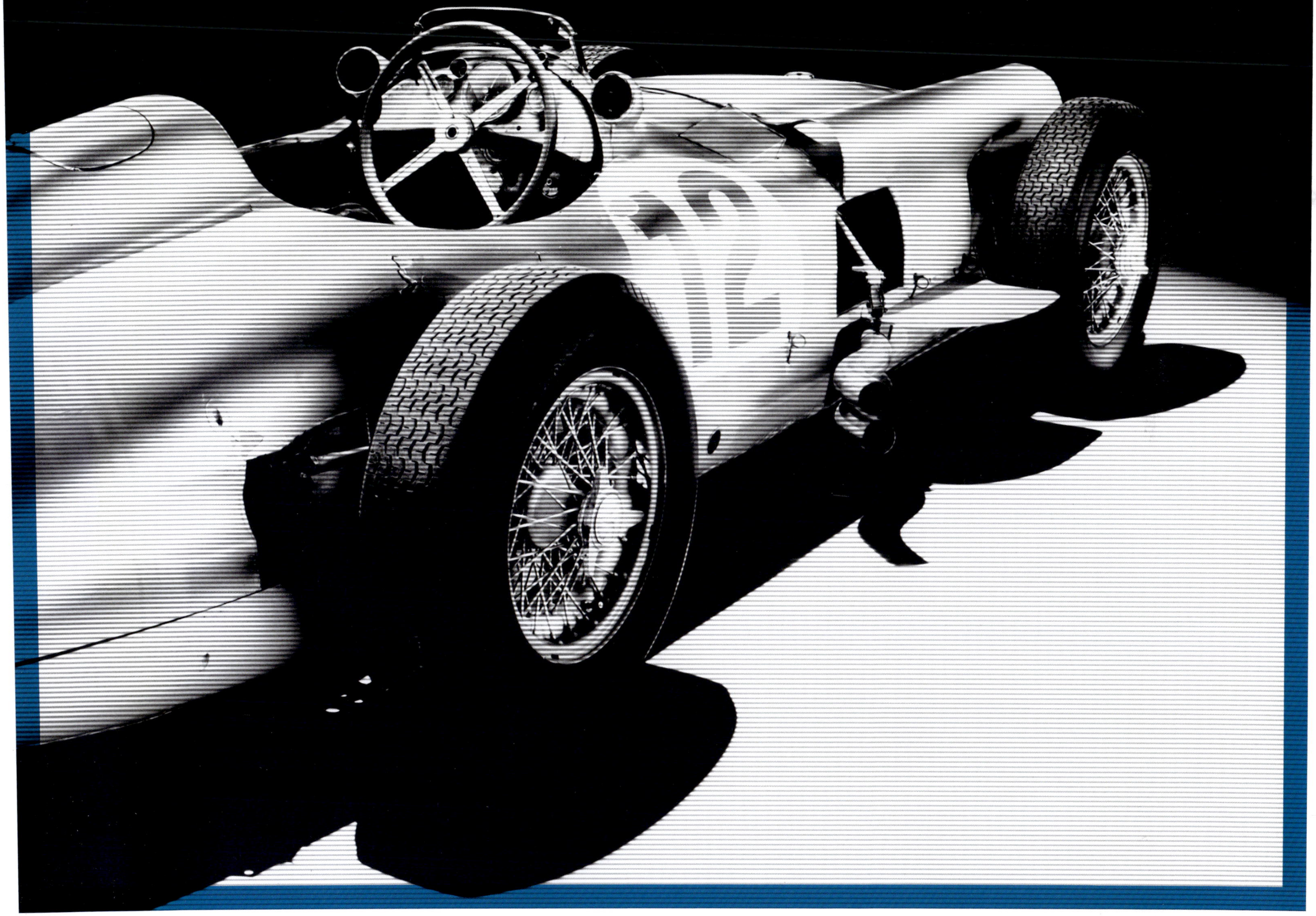

1

S-KLASSE
S-CLASS

FASZINATION ERLEBEN: DIE WELT VON MERCEDES-BENZ CLASSIC.

HISTORY THAT CAPTIVATES: THE WORLD OF MERCEDES-BENZ CLASSIC.

Klassische Automobile strahlen eine Faszination aus, die direkt ins Herz geht. Ob sie oder er, ob jung oder alt: glänzender Chrom zaubert glänzende Augen in die Gesichter der Besitzer und Betrachter.
Mit Mercedes-Benz Classic Life halten Sie jetzt ein Buch in den Händen, das mit unterhaltsamen Geschichten von ausgewählten Ereignissen über die Faszination klassischer Mercedes-Benz Fahrzeuge erzählt. Zukunft braucht Herkunft. Ohne die Erfahrung aus über 128 Jahren Automobilbau wäre die neue S-Klasse W 222 kaum denkbar. Die Baureihe ist das Aushängeschild von Mercedes-Benz und Schrittmacher der Automobilinnovation. Ihren Ursprung hat die S-Klasse 1903, als Wilhelm Maybach mit dem Mercedes-Simplex 60 PS ein Fahrzeug konstruiert, das auch als luxuriöse Reiselimousine karossiert wird. Damit war ein Automobil der Luxusklasse aus der Taufe gehoben: Ein Fahrzeug für Kaiser und Könige, für Industrielle und Prominente, die einen bis dahin nie dagewesen Reisekomfort erfahren durften. Die Gene dieses Automobils tragen die nachfolgenden Generationen der Luxus- und Oberklasse der Marke.
Ein Mercedes-Benz W 100 ebenso wie der W 116 - die Baureihe, die erstmals S-Klasse genannt wurde und noch heute mit ihren Qualitäten überzeugt. Das ist sicher auch ein Grund, warum die Braureihe bei den Automobilisten ein heiß begehrter Youngtimer ist. Doch es sind auch ein „Strich-Acht“ oder ein „Baby-Benz“, die sich in der Gegenwart als Meilensteine ihrer Jahrgänge einer hohen Beliebtheit erfreuen und dem Lebensstil seiner Besitzer Ausdruck verleihen. Der Youngtimertrend markiert eine spannende Entwicklung, in der das Auto vom Alltagsgegenstand zur Freizeitbeschäftigung seiner Besitzer reift. Damit springt auch die Faszination Mercedes-Benz über auf eine Generation von Fahrzeugen, die die Klassiker von morgen sind. Neben den Autos sind es die Menschen und Persönlichkeiten, die bei Mercedes-Benz stellvertretend für Entwicklungen oder Epochen stehen: Bruno Sacco für Fahrzeugdesign, Béla Barényi für Fahrzeugsicherheit oder Sir Stirling Moss für die Mille Miglia sind Beispiele dafür. Vielleicht hätte es diese herausragenden Leistungen aber nie gegeben, wenn nicht zuvor eine Tochter ihrer Familie einen Besuch abgestattet hätte: Mit der Reise vom heimatlichen Ladenburg zur Mutter nach Pforzheim 1888 hat Bertha Benz der Welt bewiesen, dass der Benz Patent Motorwagen zuverlässig und für jedermann und -frau bedienbar ist. Vielleicht war diese erste Langstreckenfahrt auch Inspiration für die erste Automobilwettfahrt der Geschichte, die sechs Jahre später 1894 von Paris nach Rouen stattfand. Sieger waren die Modelle von Peugeot und Panhard & Levassor – beide mit einem Antrieb „Moteur système Daimler“ aus Bad Cannstatt.

Die Faszination der Marke Mercedes-Benz reicht zurück auf die Erfindung des Automobils und der Geburtsstunde des Motorsports vor 120 Jahren – eine Welt, in die einzutauchen sich lohnt!

Viel Freude beim Lesen wünscht Ihnen,

Michael Bock
Leitung Mercedes-Benz Classic

Classic automobiles radiate a certain magic that captures your heart. One glimpse of the shining chrome and your eyes light up, no matter whether you're young or old, male or female. This book is dedicated to the magic of classic Mercedes-Benz automobiles. It's packed with entertaining tales and recollections of key events in Mercedes-Benz and motorsport history.

Every future needs a past, and the new S-Class W 222 would be inconceivable without the over 120 years of automotive development that precede it. Not only is the S-Class the standard-bearer for the Mercedes-Benz brand, it has always set the pace of innovation in the automotive world. The S-Class first came into being in 1903 when Wilhelm Maybach designed the Mercedes Simplex 60 hp, which was also available as a touring saloon. It was the birth of a luxury automobile: a vehicle for kings, emperors, industrialists and celebrities who wanted to travel in a car offering unrivalled comfort. The S-Class inheritance was passed on from one generation to the next, and can be seen in the W 100 just as much as in the W 116 – the first series to bear the S-Class name and a modern classic celebrated by automobile enthusiasts. The "/8" and "Baby Benz" are also seen as milestones of their respective eras, cars that truly reflected their drivers' lifestyles. The increasing enthusiasm today for modern classics and retro cars is an exciting development. It reminds us that cars are more than just everyday tools – they're a hobby, a passion, a way to spend your leisure time. Now, a whole generation of Mercedes-Benz cars are becoming the classics of tomorrow.

As well as legendary cars, many famous names have also become synonymous with key developments and epochs in Mercedes-Benz history: Bruno Sacco and vehicle design, Béla Barényi and vehicle safety, and Sir Stirling Moss and the Mille Miglia, just to name a few. But their tremendous achievements might never have been possible if Bertha Benz hadn't famously driven from Ladenburg to Pforzheim to visit her mother in 1888. That journey proved to the world that the Benz automobile patent was practical and reliable, and that cars could be driven by everyone – man and woman alike. It's quite possible that this first long-distance car journey inspired the world's first car race, the 1894 Paris-Rouen. Two manufacturers, Peugeot and Panhard & Levassor, shared the trophy. Both cars were powered by Daimler engines.

Mercedes-Benz's fascinating brand history goes all the way back to the invention of the automobile and the birth of motorsport 120 years ago. It's an incredible history – one that's worth exploring.

Enjoy reading,

Michael Bock
Head of Mercedes-Benz Classic

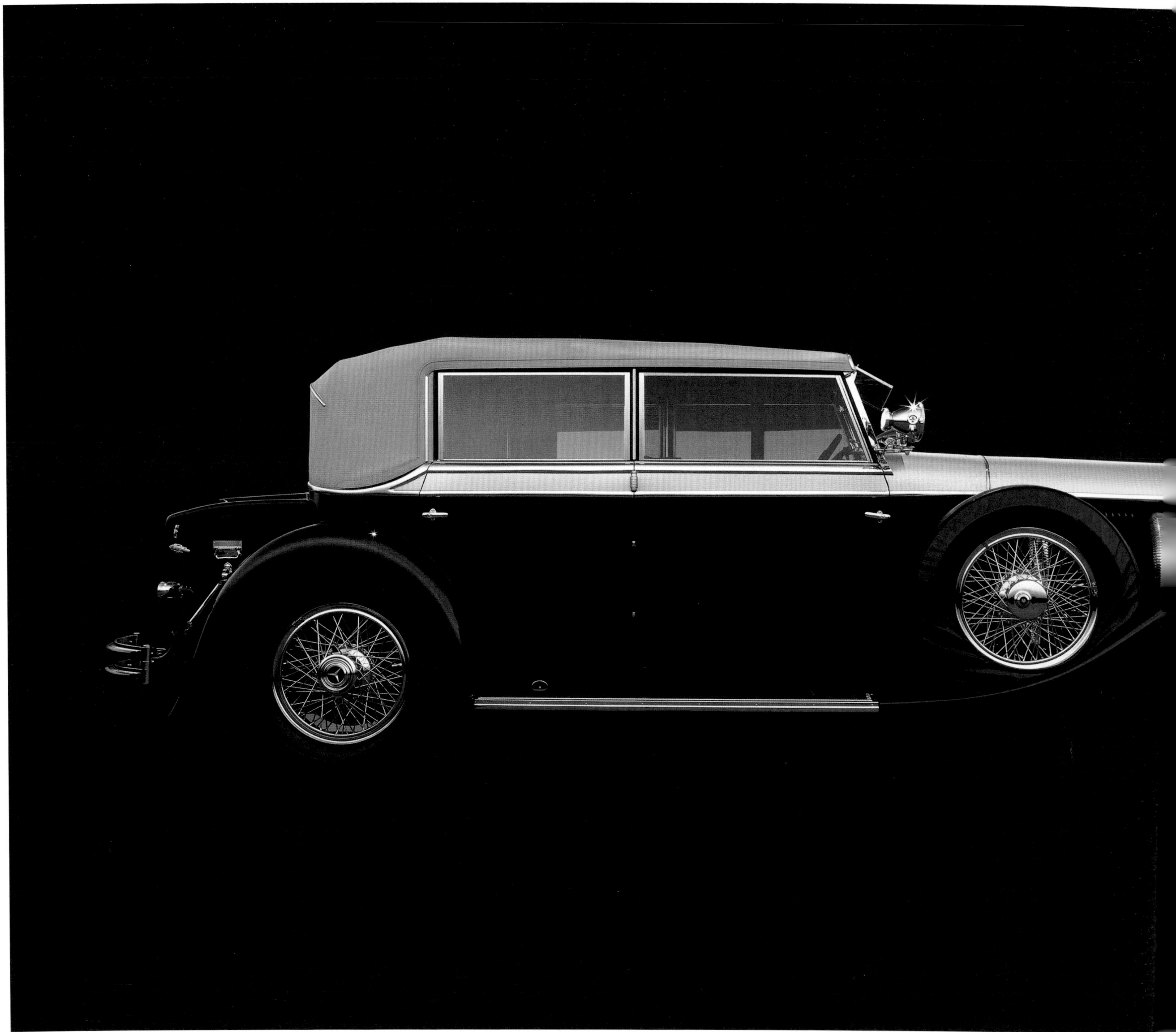

MYTHOS MERCEDES-BENZ
THE LEGENDARY MERCEDES-BENZ

DIE MARKE MIT DEM STERN –
EIN ZEICHEN FÜR INDIVIDUALITÄT.

THE MERCEDES-BENZ LEGEND
A SYMBOL OF INDIVIDUALITY.

Vorherige Seite: Ein Mercedes-Benz 770 aus dem Jahr 1931, unter Liebhabern als „Großer Mercedes“ bekannt.
Previous page: A Mercedes-Benz 770 from 1931, known to fans as the “Grand Mercedes”.

WAS MACHT DIE MARKE MERCEDES-BENZ ZUM MYTHOS?

Faszination. Perfektion. Verantwortung. Drei herausragende Merkmale, prägnant wie die drei Zacken des Wahrzeichens, das schon immer die Einzigartigkeit der Marke Mercedes-Benz unterstrichen hat. Er ist das erste, woran die Menschen zunächst denken, wenn sie den Namen Mercdes-Benz hören. Und ihm begegnen schon seit jeher Generationen von Automobilliebhabern mit Neugierde, Interesse, Leidenschaft oder sogar mit Ehrfurcht. Für Fahrer ist ihr „Mercedes“ das Spiegelbild ihrer ganz persönlichen automobilen Lebensart und die Marke Mercedes-Benz das Sinnbild einer außergewöhnlichen, traditionsreichen Automobilkultur, die Menschen auf der ganzen Welt mit einem Anspruch an das Besondere verbindet. Der Stern ist also nicht nur ein Logo, sondern vielmehr das Symbol eines einzigartigen automobilen Mythos. Er verkörpert Sicherheit. Er bietet Luxus und strahlt Erfolg aus. Was aber fasziniert so viele Menschen an Mercedes-Benz? Was macht die Marke zum Mythos?
Fragen, die unweigerlich zusammengehören, die man aber nicht einfach nur so rein sachlich beantworten will und kann. Um diesen Mythos und den kleinen Kosmos, der ihn umgibt, wirklich greifbar zu machen und verstehen zu können, sollte man sich ihm sehr real und emotional nähern. Am besten dort, wo man die Fahrzeuge mit dem Stern als Wahrzeichen antrifft: in freier Wildbahn. Auf der Straße.
Wer kennt das nicht? An der Kreuzung stehend und darauf wartend, dass die Ampel auf Grün springt, wandern unsere Blicke über den Rückspiegel nach hinten, nach links, über die Motorhaube, bis hinüber zur anderen Straßenseite. Und wir wollen dabei nicht nur weitläufig scannen, wer denn da den Bürgersteig entlangschlendert oder die Straße überquert – das natürlich auch, aber eben nicht nur – uns interessiert vor allem von welchen Autos wir umgeben sind. Wer dort drin das Steuer in der Hand hält, den Beifahrersitz belegt oder die Rückbank bevölkert. Wir wollen in den Minuten des kurz getakteten Wartens an den Ampeln, die auch schon einmal eine kleine Ewigkeit dauern können, aussagekräftige Statements empfangen und unsere eigenen aussenden.

WHAT MAKES THE MERCEDES-BENZ BRAND A LEGEND?

Fascination. Perfection. Accountability. Three outstanding characteristics as clear-cut as the three points on the Mercedes star, the logo which has always called attention to the unique nature of the Mercedes-Benz brand. The star is the first thing most people think of when they hear the name, and generations of car lovers have regarded the brand with curiosity, interest, passion or even reverence. For everyone who drives it, their “Mercedes” is a reflection of their own personal automotive lifestyle, and the brand itself embodies an exceptional automotive culture that is rich in tradition. It links people all over the world with a standard that is remarkable. In other words, the star is more than a logo: instead, it is the symbol of a unique legend. It personifies safety. It offers luxury and radiates success. But what is it that fascinates so many people about Mercedes-Benz? What makes this brand a legend? These two questions indubitably belong together, but there is no way a purely objective answer can be given, nor do we want to try and find one. If you want to fully experience this legend and the cosmos which surrounds it, want to truly understand it, you should approach the car honestly and emotionally. Ideally speaking, the encounter should happen right where you come across these cars with the star – out in the open. On the road.

Hasn’t this happened to you too? Whilst waiting at the intersection for the light to turn green, you glance at the rear-view mirror, and your gaze wanders to the left and right, to the cars ahead and behind, across the bonnet and to the other side of the street. In doing this, we not only want a wide view of who is strolling down the pavement or going over the road – that’s interesting too, of course, but it’s not everything. Above all, and most intensely, we are interested in what cars are around us. We want to know whose hands are on the steering wheel, who is sitting in the passenger seat up front or lounging around in the back seat. In the short intervals of waiting at the light, minutes which can feel like hours, we want to receive powerful statements and send out some of our own. It’s as if messages are being sent across the asphalt. This is a kind of ritual which always takes the same form.

Urbanes Asphaltgeflüster! Eine Art Ritual, immer gleich ablaufend. Menschen überall in der Welt praktizieren es. Bewusst. Unbewusst. Aber immer unabhängig davon, wo sie sich befinden, also egal ob sie in New York, London, Peking, Moskau, Berlin oder anderswo unterwegs sind. Eine nonverbale Kommunikation par excellence, eine subtile, global verständliche Sprache, die generationsübergreifend automobile Sehnsüchte vermittelt und dadurch verbindet.
Unterstützt werden wir dabei einerseits vom Motor unseres Fahrzeugs, der im Stand mit gezügelter Kraft akustisch unsere Präsenz untermauert. Gerade von jenem Motorengeräusch strömt unterschwellig eine Neugier weckende Botschaft aus, die bei uns Empfängern ganz offensichtlich eine gespannte Erwartungshaltung hervorruft. Denn wir erwarten Kraft, Persönlichkeit, Auftreten. Es scheint ein wenig so, als würden sich die Motoren schon im Standgeräusch gegenseitig zuflüstern, wer man ist und vor allem was. In unserem Fall würde man das als Sterngeflüster bezeichnen. Andererseits unterstreicht das optische Auftreten des Autos unser Dasein. Allen voran der Stern, der erhaben auf der Motorhaube thront oder unterhalb im Kühlergrill sitzt und allen zeigt, was hier vor ihnen steht, mehr noch, wen sie hier vor sich haben: einen Mercedes-Benz. Ein Statement.

Was hat das aber mit einem Mythos zu tun? Joseph Campbell, der berühmte Mythenforscher, formulierte seinerzeit einen weisen Satz mit einer klaren Aussage, aber auch einer schier unfassbaren Tiefe: „The privilege of a lifetime is being who you are". Damit brachte er eine Wahrheit auf den Punkt. Wir sind wer wir sind, ein Leben lang. Ist das aber wirklich ein Privileg? Oder eher eine Last? Oder etwa eine große Chance? Fest steht: Wir spiegeln gerne, wer wir sind und auch wer oder was wir sein wollen. Wir haben einen Ausdruck, eine Botschaft, die wir vermitteln wollen, ein Gefühl, das wir leben wollen: Individualität! Und wie kann man Individualität ein besseres Gesicht geben als mit einer eigenen Note, einer individuellen Einstellung, einem Statement, das für alle greifbar ist.
Zurück zur Kreuzung. Zum Warten. Im Auto.

DER STERN SYMBOLISIERT KRAFT, PERSÖNLICHKEIT UND EXKLUSIVITÄT.

People all over the world do this, consciously or unconsciously. No matter where they might be out and about, whether in New York, London, Beijing, Moscow, Berlin or elsewhere. This is nonverbal communication par excellence, a subtle and global language which communicates cross-generational automotive longing and creates links.

As we sit, our car's engine acoustically calls attention to our presence, its harnessed power still audible as it idles. The very sound of the engine sends out a subliminal statement that stirs up curiosity, which generates a clear sense of anticipation in those who hear it. It leads us to expect power, personality, charisma. It seems as if motors whisper to one another even when they stand waiting at the light - they tell each other who we are and, above all, what we are. In our case, with a Mercedes-Benz Classic, you could refer to this as the sound of a star whispering. At the same time, the visual aesthetic of a car defines our individuality more than ever before, especially in our age of the automatic start/stop function. And the star is what leads the way, magnificently reigning over the bonnet or resting in the radiator grille below, showing everyone what exactly is standing before them. What's more, it shows them whom they have before them: a Mercedes-Benz. A statement.
But what does this have to do with legends? Joseph Campbell, the famous mythology scholar, once expressed a wise idea that was both a clear statement and nearly unfathomable in its profundity: "The privilege of a lifetime is being who you are." This sentiment reflects something true: we are who we are throughout our lives. But is that truly a privilege? Or more of a burden? Or perhaps a great opportunity? One thing is certain: we want to communicate not only who we are, but who or what we want to be as well. We have a message, a statement that we want to convey — a feeling we want to live: individuality! And how better to give individuality a more attractive face than to lend it a unique note, a distinctive attitude, an assertion that everyone can perceive.
But back to the intersection. To waiting. In the car. We have a strong connection with what we drive. Major luminaries such as Elizabeth Taylor, John Lennon, Elvis Presley and Coco

THE STAR REPRESENTS STRENGTH, PERSONALITY AND EXCLUSIVITY.

Vorbild zu sein verpflichtet. Seit den Anfängen des Automobils sind die Spitzenmodelle von Mercedes-Benz die bekanntesten und namhaftesten Repräsentanten der Automobilbaukunst. Mit bahnbrechenden Innovationen setzen sie Maßstäbe für die gesamte Pkw-Entwicklung. Diese Tradition lebt und wird mit der Baureihe 222 fortgesetzt, die den glanzvollen Namen S-Klasse in die Zukunft trägt.

Being a role model comes with certain obligations. Ever since the beginning of automotive history, Mercedes-Benz's top-of-the-line models have been the most widely-recognized and renowned emblems of the art of manufacturing cars. With its trailblazing innovations, the company set benchmarks for automotive development everywhere. This tradition lives on and is upheld by the 222 series, which will carry the glorious name of the S-Class into the future.

Zu unserem Fahrzeug haben wir eine ganz besondere Bindung. Es sind große Persönlichkeiten wie Elizabeth Taylor, John Lennon, Elvis Presley, Coco Chanel, die durch ihre persönliche Beziehung zu einem Mercedes-Benz zeigten, wie man mit einem Automobil seine Lebensweise unterstreicht. Und auch wir wollen in allen Bereichen des Lebens, in denen wir unterwegs sind, unsere Individualität zum Ausdruck bringen. Das Auto ist unser Begleiter. Durch seine Anwesenheit trägt es unser Gefühl, unsere persönlichen Werte und das, was wir sind oder sein wollen, nach außen. Ein gelebter Mythos. So gut wie keine andere Automobilmarke wird auf so unterschiedliche Art und Weise mystifiziert wie Mercedes-Benz. Angefangen vom Wunsch, sicher mobil zu sein, über das Verlangen, schnell ans Ziel zu kommen, bis hin zu der Forderung, „das Beste vom Besten" zu fahren - alle diese Ansprüche werden von Mercedes-Benz erfüllt.

„Classic Life" taucht ein in den vielschichtigen Mythos Mercedes-Benz. Von der Reminiszenz an Bertha Benz, die das erste Automobil der Welt, den Benz Patent-Motorwagen ihres Mannes Carl Benz, zum ersten Mal fuhr und dem Automobil damit so viel Bekanntheit einbrachte, dass es sich durchsetzen konnte. Über die Erfolge des Rennsports vergangener Tage, wie der Sensationsfahrt von Rudolf Caracciola, der einen bis heute ungebrochenen Weltrekord aufstellte, den Jahrzehnte später anknüpfenden DTM-Erfolgen des supersportlichen „Baby-Benz" („Evo II"), bis hin zu technologischen Errungenschaften, wie der besonderen Konstruktion der „Pagode" mit dem einzigartigen Dach von Paul Bracq und bahnbrechenden Sicherheitstechnologien von Béla Barényi. Sie führten dorthin, wo Mercedes-Benz einst begann und sich heute nach wie vor befindet: ganz an der Spitze der Automobilbaukunst. Gespiegelt in der aktuellen S-Klasse, dem luxuriös-eleganten Flaggschiff und globalem Vorreiter in Sachen Automobiltechnologie.

Mercedes-Benz besitzt, eingepflanzt durch die Gründerväter Carl Benz und Gottlieb Daimler, die DNA des Fortschritts. Gekoppelt mit dem Anspruch an Komfort, Luxus und

Chanel used their personal connection to a Mercedes-Benz to demonstrate how a car could represent their lifestyle. And we too want to express our individuality in every aspect of our lives when we are en route. Cars accompany us, and their presence communicates to others how we interpret our personal values and what we are or want to be. It is our own personal legend that comes to life. Hardly any other brand of automobile is as mystical in as many different ways as Mercedes-Benz. The brand upholds all of these standards – from satisfying the desire to be reliably mobile to fulfilling a wish to get where you're going quickly, all the way to wanting to drive the crème de la crème.

"Classic Life" digs deep into the multifaceted legend that is Mercedes-Benz. It starts with a reminiscence about Bertha Benz, the woman who first drove the world's first car, the Benz Patent-Motorwagen created by her husband, Carl Benz. In doing so, she helped generate publicity that allowed the car to make a name for itself. After that you can read about successful historic races, such as the sensational achievement of Rudolf Caracciola, whose world record remains unbroken; the DTM successes of the super-sporty Baby Benz, the "Evo II," which followed decades later; and last but not least, technological achievements such as the special construction of the Pagoda with its unique Paul Bracq roof and trail-blazing safety technologies by Béla Barényi. These developments can all be traced back to the place where Mercedes-Benz first began and where it still remains: at the forefront of automotive construction. And this is perfectly manifest in the current S-Class, the company's luxuriously elegant flagship and the global leader in the world of automotive technology.

Thanks to the seeds planted by its founders, Carl Benz and Gottlieb Daimler, Mercedes-Benz has progress built into its DNA. Combined with its standards for comfort, luxury and individuality, this forms the basis of the legend that is a Mercedes-Benz. And the legend is not merely "installed" – it grows. Through a truly countless number of events and experiences around the world. And the people them-

Individualität begründet sie die Basis für den Mythos der Marke. Und dieser ist nicht einfach nur „installiert“, sondern auch natürlich gewachsen. Durch schier unzählige Ereignisse und Erlebnisse auf der ganzen Welt. Und egal ob Entwickler, Designer, Rennsportler oder Privatfahrer, es sind immer die Menschen und ihre persönlichen Geschichten, die die Marke prägen. Sie sind es, die der Marke ein Gesicht geben und sie zu einem lebendigen Mythos werden lassen, dessen Herzschlag man noch über Jahrzehnte wahrnimmt.

Die klassischen Automobile von Mercedes-Benz können viele Geschichten weiter erzählen, denn sie sind besondere Persönlichkeiten mit Charakter, die viel erlebt haben. Ihr einzigartiger Charme begeistert selbst eine immer jünger werdende Zielgruppe, die auf der Suche ist nach Geschichten, nach Einzigartigkeit, dem Individuellen und Besonderen, das zu ihnen und ihrem Lebensstil passt. Sicher auch ein Grund, warum die Zielgruppe von Mercedes-Benz Classic immer jünger wird, ist ein derzeit generationsübergreifender Wertetransfer. Weltweit. Wie sich das anfühlen kann, vermittelt zum Beispiel der kalifornische Youngtimer-Spezialist J.G. Francis mit seinen Jungs. Sie holen den Puls vom Gestern ins Heute, verhelfen den Klassikern des Unternehmens zu neuem Glanz und sorgen damit auf ihre Art dafür, dass der Mythos lebendig und spannend bleibt. Aber das ist nur eine von vielen Geschichten. Stars & Stories erzählt die besten und eindrucksvollsten Erlebnisse rund um die klassischen Automobile, die den Mythos Mercedes-Benz ausmachen. •

selves – no matter whether developers, designers, race car drivers or private individuals – and their personal stories are what shape the brand. They are the ones who give the Mercedes name a face and turn it into a living star, filling it with life that can be perceived across the decades.

And classic Mercedes-Benzes can keep telling countless stories, because they are specific personalities with a well-defined character, and they have experienced a great deal. Their charm is unique and they have a spark that ignites, lighting up an increasingly youthful target group which is looking for stories as well – for something unique, individual and distinctive which suits them and their lifestyles. This is certainly another reason why the target group for Mercedes-Benz Classic is growing younger and younger. Around the world, values are being passed from generation to generation. One example of the timeless appeal of the vintage can be found in Californian modern-classic specialist J.G. Francis and his team. They transport Mercedes-Benz classics from yesteryear into modern times and give them a new lustre; their restorations help make the legend dynamic and keep it exciting. But this is only one of many examples. Mercedes-Benz Classic Life tells the best and most impressive stories about the classic cars which form the foundation of the Mercedes-Benz legend. •

DIE KLAREN LINIEN EINES MERCEDES-BENZ SCHEINEN ALLE ZUM STERN ZU FÜHREN. SEIT 1910 ZIERT ER DIE MOTORHAUBEN ALLER MERCEDES-BENZ AUTOMOBILE. SEIN ÄUSSERES HAT SICH IMMER WIEDER LEICHT VERÄNDERT, DOCH SEINE BOTSCHAFT BLEIBT DIESELBE: EIN SYMBOL FÜR INDIVIDUALITÄT.

THE CLEAR LINES OF A MERCEDES-BENZ SEEM TO LEAD EVERYONE TO THE STAR. SINCE 1910 THE SYMBOL HAS GRACED THE BONNET OF EVERY MERCEDES-BENZ. ITS APPEARANCE HAS CHANGED MANY TIMES, BUT ITS MESSAGE HAS ALWAYS REMAINED THE SAME: IT IS A SYMBOL OF INDIVIDUALITY.

S-KLASSE

DER BESTE MERCEDES-BENZ DER WELT

S-CLASS

THE BEST MERCEDES-BENZ IN THE WORLD

Neuer Patent-Motorwagen

mit Gasbetrieb durch Benzin

von

Benz & Cie., Rheinische Gasmotorenfabrik

in

Mannheim.

Höchste Auszeichnung (Ehrendiplom) Ausstellung Glogau 1888.

Ausgestellt in der Kraft- und Arbeitsmaschinen-Ausstellung in München

seit dem 12. September 1888.

Die Münchener Blätter schreiben über diesen Wagen:

Neueste Nachrichten, 16. September:

In der Kraft- und Arbeitsmaschinen-Ausstellung finden seit gestern mehrere kurze Probefahrten mit dem patentirten Motorwagen der Firma Benz & Co. in Mannheim statt. Der Wagen ist eine viersitzige komfortable Kalesche mit nmlegbarem Regendach und ruht auf drei Rädern, deren vorderes durch eine äußerst einfache Vorrichtung, ähnlich wie bei den Tricycles gesteuert werden kann. Hinter dem Rücksitze im Wagenkasten ist ein kleiner zierlich gearbeiteter Benzinmotor angebracht, der mit den Hinterrädern in Verbindung steht. Durch einen Druck auf einen unter der Steuervorrichtung angebrachten Hebel setzt sich der Wagen in Bewegung. Derselbe kann auf verhältnißmäßig kleiner Fläche umkehren und sofort zum Stehen gebracht werden. Seine größte Fahrschnelligkeit ist die eines in scharfem Trabe fahrenden Fuhrwerkes.

Neueste Nachrichten, 18. September:

Der Benz'sche Patent-Motorwagen, der in der Kraft- und Arbeitsmaschinen-Ausstellung ausgestellt ist, unternahm am Samstag seine erste Probefahrt in den Straßen der Stadt und zwar mit vorzüglichem Erfolg. Das seltsame Fahrzeug wurde allgemein angestaunt und die liebe Jugend verfolgte es in dichten Schaaren. Von nun ab finden täglich von 2 bis 4 Uhr Probefahrten in den Straßen statt.

BERTHAS BENZ

UND DER TRAUM DER PFERDELOSEN KUTSCHE.

AND THE DREAM OF THE HORSELESS CARRIAGE.

Der Benz Patent-Motor trieb den am 29. Januar 1886 patentierten Motorwagen an.
The motorcar patented on 29 January 1886 was propelled by Benz's ingenious motor.

Mut, Leidenschaft und ein außergewöhnliches Gespür für die Anforderungen und Chancen der Zukunft, für echte Avantgarde im technischen wie auch gesellschaftlichen Sinne. So könnte man die emotionale DNA von Mercedes-Benz beschreiben. Begründet ist sie auf dem außergewöhnlichen, ja für ihre Zeit geradezu rebellischen Einsatz einer ungewöhnlichen, klugen Frau: Bertha Benz. Manchmal ist es erstaunlich, wie sich Geschichte wiederholt, in einem moderneren Gewand zwar, aber basierend auf verblüffend ähnlichen Gedanken. Als Bertha Benz vor nunmehr über 125 Jahren heimlich zu ihrer legendären Fahrt aufbrach, um die 106 Kilometer von Mannheim nach Pforzheim – die erste Fernfahrt mit einem Automobil – zurückzulegen, war das eine Weltpremiere. Und der Durchbruch für den pferdelosen Wagen, der bis dahin von seinen zahlreichen Kritikern belächelt und geschmäht wurde. Ohne den starken Willen und den unerschütterlichen Erfolgsglauben seiner Frau, hätte Automobilpionier Carl Benz seine geniale Erfindung, das erste Automobil, niemals so wegweisend an den Start gebracht. Doch Bertha Benz war fest davon überzeugt: Der selbstfahrende Wagen wird die Zukunft, man muss die Kritiker und Spötter nur von seiner Leistungsfähigkeit überzeugen.
Diese Gedanken kommen Ihnen irgendwie bekannt vor, als hätte man gerade gestern darüber diskutiert? Hat man auch. In diesem Jahr hat Mercedes-Benz auf der historischen Route, die Bertha Benz damals wählte, wieder eine spektakuläre Premiere gefeiert, diesmal mit einer autonom fahrenden S-Klasse der Baureihe 222. Und wieder hat das lange unmöglich Geglaubte funktioniert. Und noch viel besser, als man es sich lange hatte vorstellen können. Wieder hat Mercedes-Benz die Konkurrenz abgehängt, fährt sicher in die Zukunft, während andere noch üben.

Doch zurück zu Bertha Benz, mit der alles begann und die mit sicherem Gespür dafür sorgte, dass ihrem Mann der Erfolg zuteil wurde, den er verdiente. Das begann bereits schon vor ihrer legendären Fahrt, während der Verlobungszeit, in der Carl Benz durch seinen Geschäftspartner August Ritter in eine nahezu ausweglose wirtschaftliche Situation geriet, die Bertha Benz resolut entschärfte, indem sie sich ihre Mitgift vorzeitig auszahlen ließ. Völlig sicher in ihrem Gefühl, dass es keine bessere Anlage geben könne als die Firma ihres Mannes.

Geschafft war es aber noch lange nicht. Es kamen weitere Rückschläge. Sogar noch, als das Gefährt längst einige Kilometer am Stück ohne größere Zwischenfälle zurücklegte, und das mit einer Spitzengeschwindigkeit von bis zu 20 km/h, und Benz anschließend am 29. Januar 1886 stolz seinen Motorwagen zum Patent anmeldet. Benz war noch immer weit

If you had to describe Mercedes-Benz's emotional DNA, it could be defined as consisting of courage, passion and an unusual awareness of future needs and opportunities – of the true avant garde in both the technical and social sense. It has its roots in the extraordinary efforts of an exceptional, intelligent woman, efforts that were downright rebellious for their time: Bertha Benz.
At times it is remarkable how history repeats itself – events are sometimes inspired by unbelievably similar ideas, even if they do crop up dressed in modern clothing the second time around. Over 125 years ago Bertha Benz made history with her world premiere when she secretly struck out on her now-legendary trip of 106 kilometres from Mannheim to Pforzheim. It was the first-ever long-distance trip by car. It was also a break-through for the horseless carriage, whose numerous critics smirked and ridiculed it. Without his wife's resolute will and unshakable confidence in his success, automotive pioneer Carl Benz would never have managed to launch his brilliant invention – the first automobile – in such a trailblazing manner. Bertha Benz, however, was solidly convinced that the self-propelled vehicle was the path to the future, if only the hecklers and cynics could see how powerful it was.

Does something about this story sound especially familiar? Almost as if you had heard someone talking about it just yesterday? Because that is exactly what happened. In 2013, Mercedes-Benz celebrated a spectacular debut along Bertha Benz's historical route, but this time, the vehicle in question was a self-driving 222 model S-Class. And once again, a Benz achieved something which was long believed to be impossible – and with even greater success than was hoped for. Once again, competitors continued to take baby steps whilst Mercedes-Benz left them in the dust and safely travelled down new roads to the future.

Let us return, though, to Bertha Benz, who started it all and confidently saw to it that her husband received the success he deserved. Her active support began well before the legendary road trip: during her engagement to Carl Benz, his business partner August Ritter nearly brought the company to the brink of ruin, and Bertha Benz resolutely resolved the crisis by having her dowry paid out in advance. She was unwavering in her certainty that there could be no better investment than her husband's company.

The Benz family still had a long road ahead, however. There were other setbacks, even after Carl's invention had safely covered distances of several kilometres at a stretch without any incidents, reaching top speeds of up to 20 km/h. At that point Benz proudly registered the patent for his automobile on

Bertha Benz unternahm im August 1888 die erste Automobil-Fernfahrt der Welt von Mannheim nach Pforzheim.

In August 1888, Bertha Benz made the first-ever long-distance car journey, from Mannheim to Pforzheim.

Mit dem Patent-Motorwagen wird Carl Benz 1886 zum Erfinder des Automobils.
In 1886, Carl Benz invented the automobile – his Patent-Motorwagen.

[redacted]

entfernt vom großen Durchbruch. Nun war der pferdelose Wagen zwar Stadtgespräch, allerdings als „Teufelsgefährt".
Die große Masse belächelte das neuartige Gefährt, und forderte, die Straße gehöre den Pferden. Und so wollte der dreirädrige Benz Patent-Motorwagen zunächst keine Käufer finden. Wenn Bertha nicht gewesen wäre, wer weiß, vielleicht hätte Carl Benz dann resigniert. „Jetzt merkte ich, dass ich ganz unten stand, ganz unten, und wie ein Bettler anklopfen musste vor den Türen der Menschheit und ihrer Kultur", notierte er später in seinen Memoiren.
Erlaubt hätte er das Abenteuer, zu dem Bertha sich entschloss, um endlich allen zu zeigen, was wahre Innovationskraft ist, dennoch niemals. „*Er wäre zu zögerlich gewesen*", sagt Winfried A. Seidel, Leiter des Automuseums Dr. Carl Benz e.V., der die Familiengeschichte kennt, wie kaum ein anderer.
Was für eine Nacht- und Nebelaktion war das, zu der Bertha Benz, damals 39, Anfang August 1888 mit ihren beiden Söhnen Eugen und Richard aufbrach.

Gabriele Spangenberg, Autorin und Rennfahrerin aus Leidenschaft – im August wurde sie nach dem Sieg beim AvD-Oldtimer-Grandprix 2013 sogar mit der Stirling Moss Trophy für die couragierteste Fahrt ausgezeichnet – lässt das Bertha-Benz-Abenteuer lebendig werden:

„Ich stelle sie mir gerade vor, diese Fahrt: hundert Kilometer über Stock und Stein. Die ersten Straßen entstanden wohl in Preußen im letzten Drittel des 19.Jahrhunderts. Ich stelle mir vor, wie sie ihr Strumpfband, vor den Augen der prüden Öffentlichkeit versteckt, unter ihren Kleidern hervorzieht, um ein durchgebranntes Zündkabel zu isolieren. Wie sie dann mit einem rutschenden Strumpf – ziemlich ungemütlich so etwas – ganz cool weiterfährt. Wie sie eine Hutnadel aus Hut und Haaren zieht, um die verstopfte Benzinleitung frei zu stochern und diese benzinstinkende Hutnadel wieder in Hut und Haare steckt, oder ohne Hut weiterfährt. Wie sie ständig nach Apotheken Ausschau hält, um Waschbenzin in Flaschen – nein, auch Kanister gab es damals noch nicht – zu kaufen. Wie sie gleichzeitig, über den Höllenlärm des Gefährts und das Aufsehen in der Bevölkerung beunruhigt, nach Polizisten Ausschau hält. Denn schließlich hatte ihr Mann eine nur beschränkte Betriebsgenehmigung für Mannheim erhalten.
Ich wüsste gerne, was ihre Mutter gedacht hat, als Tochter und Enkel in einem sicherlich nicht sehr damenhaften, sondern ölverschmierten Aufzug bei ihr auftauchten.

Sie scherte sich nicht um Aussehen, um Würde oder sonstige gesellschaftliche Konventionen. Sie wollte nur eins: Sie wollte der Welt beweisen, was das Automobil ihres Mannes wirklich leisten konnte.
Der wusste übrigens nichts von dieser Reise – sie war heimlich bei Morgengrauen aufgebrochen. Ein Telegramm, in dem stand, sie seien gut bei Oma angekommen, klärte ihn darüber auf.
Bertha hat ihr Leben lang die Möglichkeit zu scheitern einfach weggeschoben und missachtet. Sie ignorierte Widerstände, ließ sich nicht abschrecken und schon gar nicht aufhalten. Ein Freigeist, eine Heldin. Ich bin begeistert von dieser Frau, die sich vor über 125 Jahren unter ungeheuer großen Beschränkungen behauptete. Wie aber sähe ihr Leben heute aus? Was würde sie heute tun? Die Entwicklung

29 January 1886. But he was still far from achieving his goal: the horseless carriage was indeed the talk of the town, but it was regarded as "the devil's vehicle".

The public at large mocked the newfangled conveyance, insisting that the streets were meant for horses. Subsequently, the three-wheeled patented Benz motorcar had difficulty finding customers.
If it were not for Bertha, Carl Benz might have given up. "Now I saw that I was at the very bottom, the very bottom, and like a beggar I had to knock at the doors of humanity and its culture", he later noted in his memoirs. Nevertheless, he would never have permitted Bertha to set out on the adventure she set her mind to in her effort to show everyone what true innovative power is. "*He would have been too reluctant*", according to Winfried A. Seidel, the head of the Automuseum Dr. Carl Benz, whose knowledge of the family history is almost peerless. What a cloak-and-dagger operation it was when 39-year-old Bertha Benz headed out with her two sons Eugen and Richard in early August 1888!

Gabriele Spangenberg is an author and passionate race-car driver who won the AvD Oldtimer Grandprix 2013 in August and was even awarded the Stirling Moss Trophy for being the most courageous driver. She vividly describes Bertha Benz's adventure:

"I like to imagine her trip – a hundred kilometres over hill and dale. The first streets were not built until the last third of the 19th century in Prussia. I imagine her reaching under her skirt and taking off the garter which was safely concealed from the eyes of the prudish public so she could insulate an ignition cable. And then she kept driving as cool as a cucumber even as her stockings kept slipping, which is rather unpleasant. I think about her taking a hairpin from her hat to unplug the clogged-up fuel line and then putting the reeking hairpin back in her hat, or even driving without her hat. About the way she constantly kept a lookout for pharmacies so she could buy more bottles of ligroin for fuel – and no, there was no such thing as canisters at the time. The way she also stayed vigilant so as to avoid the police over the roaring noise of her vehicle and the crowds she was drawing. After all, her husband only had limited permission to drive in Mannheim.

I would love to know what her mother must have thought when her daughter turned up without warning, looking quite unladylike in her oil-smeared clothes and with her sons in tow. She could not be bothered with appearances, dignity or other societal conventions. She only wanted one thing: to show the world what her husband's car was capable of. He, by the way, knew nothing of her journey: she had simply headed out at daybreak. He learned of it when he received a telegramme informing him that they had arrived at her mother's house.

It's crazy. All her life, Bertha must have simply pushed aside and actively ignored the possibility of failure. She paid no heed to obstacles and was not discouraged, let alone deterred, by challenges. She was a free spirit, a hero. I find this woman delightful: over 125 years ago she asserted herself in the face of unbelievable opposition. What would her life be like if she were around today? What would she do? Support the development of hydrogen-powered cars and take part in Le Mans?

Auf der ersten Automobil-Fernfahrt der Welt 1888 sorgten Apotheken für den Kraftstoffnachschub.
During the world's first long-distance drive in 1888, Bertha Benz had to refuel at pharmacies.

wasserstoffbetriebener Autos unterstützen und in Le Mans teilnehmen? Das Elektroauto vorantreiben? Batterien entwickeln? Alternative Antriebskonzepte auf den Weg bringen?

Eins ist sicher: sie würde sicher auch heute noch auf ihr Bauchgefühl hören, sie würde ihr Leben mit Hingabe und Passion führen, eine Aufgabe suchen, die Visionen, außergewöhnlichen Mut und Unabhängigkeit von Status und Geld erfordert. Und Bertha Benz hätte ganz sicher große Freude daran gehabt, dass es Mercedes-Benz war, der als erster Automobilbauer ein autonom fahrendes Fahrzeug 100 Kilometer fahrerfrei absolvieren ließ – auf einer Strecke, auf der einst eine automobile Revolution begann. Mut, Leidenschaft und ein außergewöhnliches Gespür für die Anforderungen und Chancen der Zukunft, für echte Avantgarde im technischen wie auch gesellschaftlichen Sinne. Bertha Benz darf stolz darauf sein, dass ihr Geist das Unternehmen auch heute noch befeuert und in die Zukunft trägt. •

Promote the electric car? Develop batteries? Work on alternative fuel concepts?

One thing is certain: if she were alive today, she would certainly be listening to her gut feeling, filling her life with dedication and endless enthusiasm, and finding a cause which demanded vision, exceptional courage and independence from status and wealth."
And Bertha Benz would certainly have been thrilled that Mercedes-Benz was the first car manufacturer to put a self-driving car on the road. It successfully covered a distance of 100 kilometres – on the very same route where the automobile revolution started. Courage, passion and an unusual awareness of the needs and opportunities of the future – for the true avantgarde in both the technical and social sense. Bertha Benz would be proud that her spirit is what continues to inspire and drive the Mercedes-Benz brand forward to this day, spurring it on to tomorrow. •

Früher war die Route die Herausforderung. Heute ist es die Technik. Trotzdem sind die Eckdaten der autonomen Fahrt spektakulär.
The route itself was once the real challenge, whereas nowadays it is technology. But the key facts and figures about self-driving cars are spectacular, no matter what.

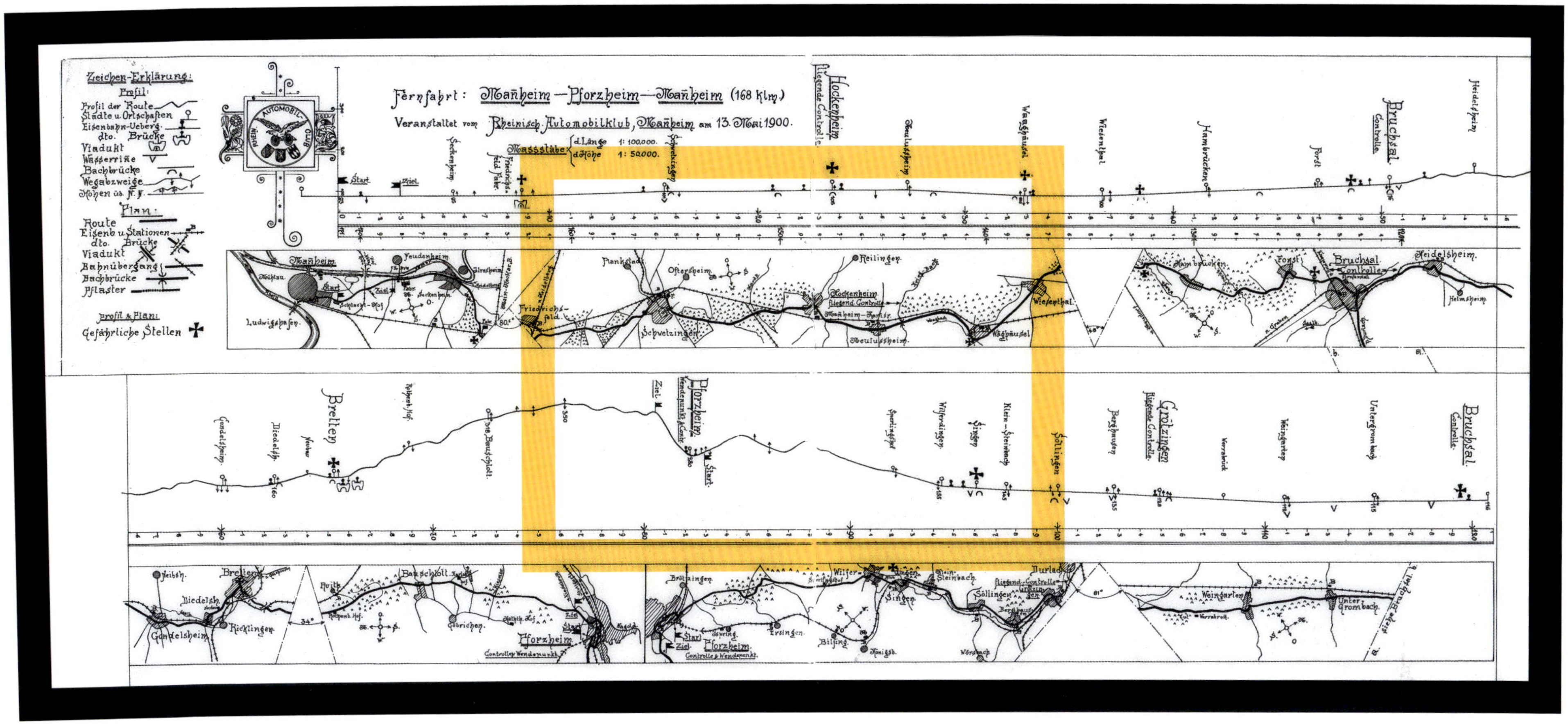

AUTONOMES FAHREN AUF DEN SPUREN DER BERTHA BENZ

103 KILOMETER GESAMTSTRECKE
50 KILOMETER ÜBERLAND
53 KILOMETER STADT
18 KREISVERKEHRE

A SELF-DRIVING CAR ALONG BERTHA BENZ'S FAMOUS ROUTE

103 KM TOTAL DISTANCE
50 KM CROSS-COUNTRY
53 KM IN URBAN AREAS
18 ROUNDABOUTS

1
Auf historischem Boden: Das Forschungsfahrzeug S 500 INTELLIGENT DRIVE.
On historic ground: the S 500 Intelligent Drive test vehicle.

2
Auf den Spuren von Bertha Benz geht es von Mannheim nach Pforzheim.
Following Bertha Benz‘s footsteps from Mannheim to Pforzheim.

3
Getestet wird das vollkommen autonome Fahren im Überland- und Stadtverkehr.
The fully autonomous car was tested both in city traffic and cross-country.

4
Dank eines ausgeklügelten Systems an Spezialtechnologien...
Thanks to a sophisticated system of special technologies...

5
... und eines perfekt organisierten Ablaufplans...
... and a perfectly organised schedule...

6
... kommt der S 500 INTELLIGENT DRIVE selbstständig in Pforzheim an.
... the S 500 Intelligent Drive makes its way to Pforzheim all by itself.

7
Ein Dutzend Nah- und Fernbereichsradare überprüfen den Fahrkorridor.
A dozen short-range and long-distance radars monitor the route.

8
In Mannheim startete 1888 auch Bertha Benz ihre erste große Fahrt.
Bertha Benz struck out on her first major trip from Mannheim in 1888.

9
Heutzutage findet und überprüft das Fahrzeug seine Strecke eigenständig.
Nowadays the vehicle can find and monitor its route independently.

10
Und meistert dabei auch hochkomplexe Situationen wie Fußgängerüberwege.
And can even master highly complex situations such as pedestrian crossings.

11
Über ein Display kann der Fahrer alle Handlungen des Fahrzeugs mitverfolgen.
A display lets the driver follow everything the car is doing.

12
Auch Ampeln oder Kreisverkehre stellen so kein Problem dar.
Even traffic lights and roundabouts are not a problem.

[redacted]

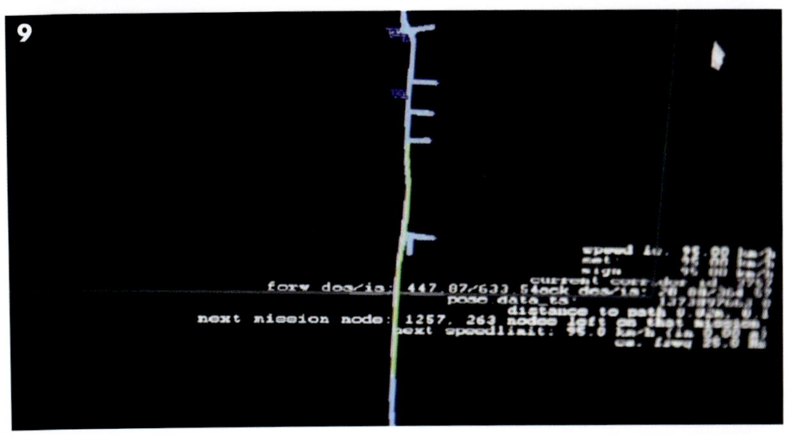

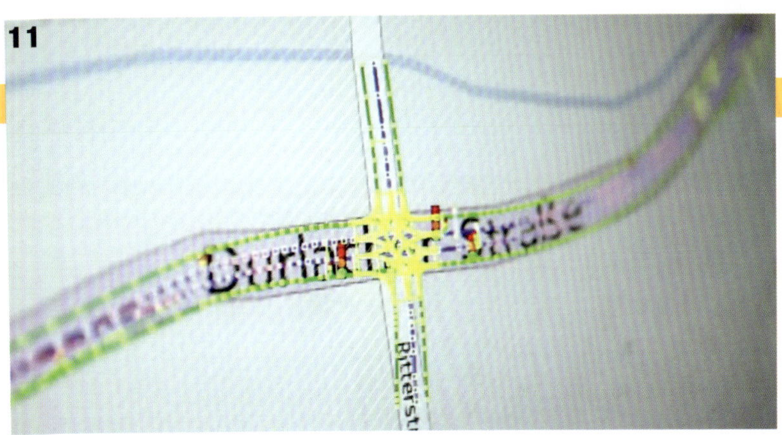

300 SE
BB-C 540

1.2

DIE GESCHICHTE DER S-KLASSE

THE HISTORY OF THE S-CLASS

TRADITION VERPFLICHTET – DIE S-KLASSE WAR SCHON IMMER STATE OF THE ART.

TRADITION OBLIGES – THE S-CLASS HAS ALWAYS BEEN STATE OF THE ART.

Ab 1972 gibt es einen neuen Namen für die berühmten Oberklasselimousinen von Mercedes-Benz: S-Klasse.
From 1972 onwards, Mercedes-Benz's renowned premium range of cars bears a new name: the S-Class.

[redacted]

Haben Sie schon einmal einem Auto richtig bewusst in die Augen geschaut? Ist es nicht so, als würde man in ein ausdruckstarkes Gesicht blicken, eines, das sofort zurück blickt und einen dabei ganz genau spüren lässt, wen man da eigentlich vor sich hat? Die Mercedes-Benz S-Klasse ist viel mehr als nur ausdrucksstark. Sie hat dieses eine, ganz be-sondere Etwas, einen Anspruch, mit dem sie schon immer zu überzeugen wusste.

Reiht man alle Generationen von S-Klassen im Vergleich aneinander, fällt auf, dass sich dieser Anspruch in allen Fahrzeugen wiederspiegelt. Techniker und Ingenieure pflanzten ihn tief in das Erbgut des Mercedes-Benz Flaggschiffs. Seitdem wird er von Generation zu Generation weitergereicht. Der Zeit angepasst, haben sich Proportionen und Linien, nie aber der grundlegende Charakter.

denen Bereichen. Sie ist also Extraklasse. Tatsächlich spricht man bei ihr von „Sonderklasse". Dafür steht auch das bekannte Kürzel „S". Was macht aber diese Sonderklasse aus? Woraus entwickelte sie sich? Was ist ihre zentrale Botschaft? Fragen, die sich auf den ersten Blick nicht ergründen lassen. Begeben wir uns auf Spurensuche in der Automobilgeschichte.

Bevor man in die Vergangenheit eintaucht, stößt man zuerst einmal auf die essentielle Frage, was bedeutete uns das Auto früher und was heute? Und ganz salopp könnte man daraufhin antworten: „Alles! Denn wir lieben Autos." Man könnte ergänzen: „Wie einst die Pferde".
Betrachtet man dann die allgemein gängige Analogie von Auto und Pferd, wird das enge Geflecht zwischen Mensch und Auto deutlich.

BEIM FAHREN BESCHWÖRT DIE S-KLASSE DESHALB SEIT JEHER EIN GANZ BESONDERES GEFÜHL HERAUF: ERHABENHEIT.

Bei der S-Klasse handelt es sich um ein ganz besonderes Automobil. Egal aus welcher Epoche es stammt. Und es macht große Freude, selbst am Steuer zu sitzen. Oder, sich chauffieren zu lassen und dabei inmitten der neuesten Technologie der jeweiligen Zeit zu sitzen, umgeben von edlen Materialien und manifestiert durch eine Verarbeitung, die den Begriff „Made in Germany" prägt wie sonst fast nirgendwo. Spätestens dann wird deutlich, dass die S-Klasse von Mercedes-Benz schon immer ihrer Zeit weit voraus war.
Würde man sie einer Stilrichtung zuordnen, könnte man von Eklektizismus sprechen, vereint die S-Klasse das Beste aus verschie-

Das Pferd, einst wichtigstes Fortbewegungsmittel, war oft viel mehr als nur die vorantreibende Pferdestärke, es war nicht selten Wegbegleiter oder sogar enger Freund des Menschen. Das Automobil hat es von der Seite des Menschen verdrängt, seine ursprüngliche Stellung eingenommen, ist nun selbst das Objekt, zu dem eine enge, soziale Bindung besteht und das gepflegt, verhätschelt und geliebt werden will. So weit so gut zur emotionalen Bindung Mensch-Auto, aber der Vergleich zwischen Pferd und S-Klasse birgt noch einen weiteren Vergleich, der über das Beziehungsgeflecht Mensch-Pferd hinaus geht. So ist die S-Klasse auch als Nachfolger der

Have you ever looked at a car head-on and consciously gazed into its "eyes"? Doesn't it remind you of looking at an expressive face, at someone who returns your gaze without hesitation, giving you a clear sense of their real character? The Mercedes-Benz S-Class is much more than just strong outward expression. It possesses that certain "je ne sais quoi" – a quiet, confident sophistication that has always known just how to reassure you.

Place all generations of the S-Class next to one another and compare them, and you see this sophistication mirrored in every vehicle. Technicians and engineers have planted it deeply

DRIVING AN S-CLASS HAS ALWAYS CONJURED UP A SINGULAR FEELING: MAJESTY.

in the DNA of this Mercedes-Benz flagship; a sovereign assurance that is passed on from one generation to the next. Proportions and lines may shift with the times, but never the underlying character. Driving an S-Class has always been an experience to conjure up a singular feeling: majesty. Regardless of the era it may have been born into, it's immediately clear that the S-Class is a very special automobile. It is simply an unadulterated pleasure to sit behind the wheel, or to let yourself be chauffeured, surrounded by the latest technology of the era and by fine materials, all crafted with a workmanship that embodies the term "Made in Germany" as virtually no-one else can. By this point, it is clear: the Mercedes-Benz S-Class was always ahead of its time.

If you tried to attribute this vehicle to a specific period of art, you could speak of eclecticism, as the S-Class seamlessly integrates the best from many different areas. It occupies an extra class of its own – a "special class". In fact, that's precisely what the "S" abbreviation stands for. But what is it that actually makes these cars a special class? What is their origin? What message do they convey? The answers to these questions cannot be divined at first sight. For this, we must sift for further clues in the history of the automobile.

Before embarking on our excursion into the past, we naturally first encounter one essential question: What did the car mean to us in the past, and what does it mean to us now?

One could, of course, casually answer "Well, everything! We just love our cars." And one also might add: "Just as we once loved our horses." If we take the analogy commonly drawn between automobiles and horses, the intimacy of the interaction between person and automobile becomes clear. Once the most important means of transport, the horse meant much more than just horsepower. For the rider, it was often a travelling companion and even a close friend. Today, the automobile has driven the horse away and taken its place at mankind's side. Just as the horse was, it is an object to which we have a close social connection; one that we care for, pamper and love. Beyond the emotional connection between humans and their automobiles, the horse and rider comparison holds another parallel, one that transcends the personal connection between driver and car. Though this lens, the S-Class is the successor of the majestic car-

prächtigen Karossen zu sehen, die hinter großen Pferdegespannen hergezogen wurden. Eine Art Domizil auf Rädern, das Schutz und Geborgenheit bietet, Funktionalität und eine Menge Komfort. Ihre wichtigste Aufgabe aber: Sie spiegelt den Anspruch ihres Besitzers wieder. Sie repräsentiert!

Die Ursprünge der heutigen S-Klasse waren rau und abenteuerlich. Ihre Wurzeln reichen weit bis in die Frühzeit des Automobils zurück, bis ins Jahr 1901. Deutschland erlebt gerade die Hochzeit der industriellen Revolution, und der Motorsport erlebt seine wilden, meist noch ungeordneten Anfänge.
Damals entfacht der noch junge Rennsport die Leidenschaft des österreichisch-ungarischen Geschäftsmannes Emil Jellinek für das Automobil. Dieser ist nicht nur begeisterter Zuschauer von Automobilrennen, sondern fährt sogar selbst und tritt dabei als „Monsieur Mercédès" an den Start. Ein abenteuerliches Wagnis zur damaligen Zeit. Denn jedes Rennen kann in einem tödlichen Unfall enden. Sicherheitsvorschriften gibt es noch so gut wie keine. Der Heißsporn benennt sein Auto nach dem Namen seiner Tochter: „Mercédès". Als „Monsieur Mercédès" wird er erfolgreich, macht sich einen Namen. Doch auch als Geschäftsmann brennt Jellinek vor Enthusiasmus. Er wittert ein neues, lukratives Geschäft, denn die Welt ist automobilbegeistert, das Auto in dieser Zeit das Statussymbol schlechthin. Jellinek ist überzeugt von den Fahrzeugen des Entwickler-Duos Daimler und Maybach und will künftig ihre Automobile vertreiben. Aus Erfahrung weiß er, dass es keine bessere Werbung gibt als bei großen Automobilrennen erfolgreich aufzutreten. Hier sitzt, steht und feiert das begeisterte Publikum das „technische Wunder" Automobil, hier wird mitgefiebert, entstehen Träume, keimen Wünsche, und hier werden letztendlich auch große Geschäfte gemacht. Deshalb beschließt Jellinek, die Automobile der DMG ins Rennen zu schicken. Mit Erfolg. Jellinek will mehr, drängt auf stärkere und schnellere Wagen, beeinflusst die Neuentwicklung bei der Daimler-Motoren-Gesellschaft (DMG). Ende 1900 entsteht der Mercedes 35 PS. Ein Fahrzeug, das nicht mehr aussieht wie eine Kutsche, sondern etwas komplett Neues darstellt: Es ist das erste moderne Auto. Gleichzeitig einer der ersten Sportwagen in Serie und das erste Automobil, das offiziell den Namen Mercedes trägt. Vorne sichtbar auf dem Kühler. Der heute weltberühmte Markenname ist geboren.

Jellineks Strategie geht auf. Der Mercedes 35 PS triumphiert schließlich 1901 bei der Rennwoche von Nizza. Der Geschäftsmann zieht die ersten großen Geschäfte an Land und gibt so bei der DMG die erste Großbestellung auf. Der Erfolg hält an. Die Ansprüche der Kunden verändern sich. Im Gesamten und im Individuellen. 1902 schlägt die DMG verstärkt den Weg der modularisierten Bauweise ein, gleichzeitig auch ein Schritt in Richtung Individualisierung im Automobilbau. Das weiterentwickelte Automobil nannte man schlicht Mercedes-Simplex. Mit ihm beginnt die eigentliche Ur-Geschichte der S-Klasse.

Tatsächlich war der Mercedes-Simplex der erste wirklich große technologische Meilenstein, ein sehr fortschrittliches Automobil,

DIE URSPRÜNGE DER HEUTIGEN S-KLASSE WAREN RAU UND ABENTEUERLICH.

riages that were once drawn by powerful teams of horses. A virtual home on wheels, offering protection and privacy, functionality and a considerable amount of comfort. But its most important function was another: to mirror the status and standards of the owner. To represent, to assure, and to reassure.

The origins of today's S-Class were rough and adventuresome. Its roots can be traced back to 1901, to the early origins of the automobile itself. Germany was experiencing the peak of the industrial revolution. The world of motorsports was still in its wild, largely untamed beginnings.

This young sport of motor racing ignited a passion for automobiles in Emil Jellinek, an Austro-Hungarian businessman. More than just an enthusiastic bystander, he raced himself and appeared on the starting line as "Monsieur Mercédès". This was a daring risk to take at the time, as any race could have ended in a deadly accident, and there were as good as no safety precautions. As a racing driver, "Monsieur Mercédès" – who had named his car after his daughter, Mercédès – was highly successful and made a prominent name for himself. Yet the businessman inside him was equally shrewd. The entire world had gone car mad, and the automobile was quickly turning into the ultimate status symbol. In this, he sensed the emergence of a new and lucrative business. Jellinek was a fan of the cars developed by the duo of Daimler and Maybach, and very much wanted to sell their cars in the future. From experience, he knew there was no better way to advertise than by competing in – and winning – the major races. The racetrack was where the enthusiastic public toasted the "technical wonder" of the automobile. Cars and their drivers were surrounded by adulation: fingers were crossed, dreams were born, and aspirations slowly began to bud and grow. Naturally, this was also where, at the end of the day, big business was done.
So Jellinek decided to enter racing with a car from the Daimler-Motoren-Gesellschaft (DMG). He enjoyed immediate success. And in the wake of this, he quickly began pushing for more, urging DMG to produce faster and more powerful vehicles. In the process, he exerted a strong influence on new development there. At the end of 1900, the Mercedes 35 hp was born. This automobile, which no longer looked like a horseless carriage, represented the birth of something completely new. It was the first modern car, one of the first sports cars to go into production, and the very first automobile to carry the Mercedes name, emblazoned prominently on the front cooler grille. The renowned global brand of today was born.

Jellinek's strategy quickly bore fruit. In the end, the Mercedes 35 hp triumphed at Nice Week in 1901. The businessman landed his first major customers there, and gave DMG its first big order. The successes rolled on, even as customer demands changed, both overall and individually. In 1902, DMG began pioneering the use of modular construction techniques. This also marked an important step towards customisation in automobile manufacturing. The next evolution of the original Mercedes was quite sensibly called the Mercedes Simplex, and it's with this vehicle that the pre-history of the S-Class truly begins.

THE ORIGINS OF TODAY'S S-CLASS WERE ROUGH AND ADVENTURESOME.

das Innovationen mit Luxus verband. Ein Fahrzeug, bei dem der Name Programm war, denn er besaß unter anderem eine vereinfachte Kupplung und eine verbesserte Schaltung, mit der ein einfacheres und sportlicheres Fahren möglich war. Der verlängerte Radstand verbesserte zudem die Straßenlage, und sein völlig neues Karosseriekonzept machte in variabel und individuell: Es wurden entweder Mercedes-Simplex Fahrgestelle mit Antriebsstrang ausgeliefert, was den Besitzern einen individuellen Karosseriezuschnitt ermöglichte oder komplette Fahrzeuge mit jeweils unterschiedlichen Karosserien angeboten. Das konnten sportliche Zweisitzer-Karosserien sein oder komfortabel und luxuriös ausgestattete Karosserien für Reisen. Da es noch keine automatisierte Serienfertigung gab, waren die Karosserievarianten immer etwas unterschiedlich. Dafür schuf Mercedes mit dem Simplex ein einmaliges technologisches Baukastenprinzip und in Deutschland einen Vorreiter für Individualität und Exklusivität. Retrospektiv gesehen war der Mercedes-Simplex – insbesondere die luxuriöse Reiselimousine von Emil Jellinek persönlich – der erste große Leitstern und damit der eigentliche Begründer der „S“-Klasse.

TATSÄCHLICH WAR DER MERCEDES-SIMPLEX DER ERSTE WIRKLICH GROSSE TECHNOLOGISCHE MEILENSTEIN.

Ein komprimierter Einblick in die anschließende Historie der Mercedes-Benz Oberklasselimousinen zeigt deutlich die evolutionsartige Entwicklung der Mercedes-Benz Attribute, die hin zur ersten offiziell S-Klasse genannten Baureihe führte. Auf den ersten luxuriösen Mercedes-Simplex Reisewagen mit 60 PS (44 kW) folgten über die Jahrezahlreiche Luxuslimousinen, die technisch und qualitativ immer ausgefeilter wurden. Die Ansprüche an Technik paarten sich mit Ansprüchen an Material, Verarbeitung und Haltbarkeit.
Die Techniker und Ingenieure arbeiteten voller Enthusiasmus an den Entwicklungen und identifizierten sich voll und ganz mit der Marke Mercedes-Benz. Ein Glanzpunkt der Entwicklung war etwa der ab 1930 gebaute „Große Mercedes“, der Typ 770 (interne Bezeichnung W 07), ausgestattet mit einem hubraumstarken Achtzylinder-Kompressormotor, der für souveräne Fahrleistungen bürgte. Eine weitere Stufe auf dem Weg Richtung S-Klasse, wies nach dem Zweiten Weltkrieg der Mercedes-Benz 300 (W 186) von 1951, dem Liebling des damaligen deutschen Bundeskanzlers Konrad Adenauer, der mit dessen Namen als Spitznamen in der Automobilgeschichte verewigt wurde. Der „Adenauer“ hatte einen Sechszylinder mit obenliegender Nockenwelle und wurde Ende der 50er Jahre optional sogar schon mit Klimaanlage und mit Servolenkung ausgestattet. Was kaum einer weiß: Konrad Adenauer selbst entwickelte eine Vorrichtung zur Verhinderung von Zugluft, die dann auch in Serie eingebaut wurde. Kundenwünsche im Fokus. Ein Anspruch an sich selbst, der in der Geschichte der Mercedes-Benz-S-Klasse schon immer kontinuierlich beibehalten wurde.

Den nächsten großen Schritt vollbrachten die Ingenieure 1954 mit dem Typ 220 (W 180). Er besaß eine geräumige selbsttragende „Ponton“-Karosserie und erhielt als erste Oberklassenlimousine von Mercedes-Benz ab

The Mercedes Simplex was the first major technological milestone in automobile manufacturing: an advanced vehicle that combined both innovation and luxury. As a car whose very name promised simplicity, it offered a simplified clutch and improved gear shifting that enabled an easier, sportier style of driving. The lengthened wheel base improved its road-holding and its entirely new style of construction made it both flexible and customisable. The Simplex could be delivered either as a basic chassis with a powertrain which owners could then fit with a custom body, or as a complete vehicle in several different body styles, including a sporty two-seater and a comfortable limousine for travelling.
Since there was not yet any mass production in the modern sense, each body type was slightly different. Yet with the Simplex, Mercedes created a unique new modular technology and became a German pioneer in the art of individualised, exclusive manufacturing. Viewed historically, it was the Mercedes Simplex – and in particular Emil Jellinek's own luxurious version thereof – that formed the lodestar for the birth of the S-Class.
A compressed view of the subsequent evolution of the Mercedes-Benz luxury car illustrates just how the key attributes of the brand evolved over time, finally culminating in the first official S-Class.

IN FACT, THE MERCEDES SIMPLEX WAS THE FIRST TRULY MAJOR TECHNOLOGICAL MILESTONE.

Starting with the first 59 hp (44 kW) version of the Mercedes Simplex, many other luxury cars followed, each with ever more sophisticated qualitative and technical refinements. Invariably, the latest technology was combined with discriminating standards in materials, workmanship and durability. Technicians and engineers worked passionately in the development, and identified themselves wholeheartedly with the Mercedes-Benz make. An early high point in the art of development was the Mercedes Type 770 (Grand Mercedes), known internally as the W 07. This vehicle began production in 1930 and was outfitted with a high-displacement eight-cylinder supercharged engine which guaranteed superior driving performance.
Following World War II, another significant step in the pre-evolution of the S-Class was the Mercedes-Benz 300 (W 186) of 1951. This vehicle was a personal favourite of the then German Chancellor Konrad Adenauer, who is now immortalised in automotive history with this car unofficially named after him. The “Adenauer” had a six-cylinder engine with an overhead camshaft, and by the end of the 1950s even featured optional power steering and air conditioning. What virtually no one knew is that Konrad Adenauer himself developed a mechanism to prevent air infiltration; an innovation that was later incorporated into the series. This single-minded focus on the wishes of the customer remains an in-house standard, one that has characterised the entire history of the S-Class.

Engineers made the next major breakthrough in 1954 with the Mercedes-Benz 220 (W 180), which possessed an integral “ponton” body. This was also the first luxury limousine from Mercedes-Benz to carry the “S” as a permanent part of its type designation.
In 1959, an entirely new chapter in automotive history opened: the era of the Fintail. The Mercedes-Benz 220, 220 S and 220 SE (W 111) were – and still are – characterised by a protruding rear that is simply beautiful to behold.

1956 das „S“ als dauerhaften Bestandteil der Typenbezeichnung.
1959 dann ein ganz neues Kapitel der Automobilgeschichte: die „Heckflosse“. Die Mercedes-Benz 220, 220 S und 220 SE (W 111) waren – und sind immer noch – nicht nur optisch durch ihr ausladendes Heck schön anzusehen, sie repräsentierten auch einen neuen Typus Automobil, ein Musterbeispiel für in Serie umgesetztes Sicherheitsdenken, wie es das bis dato noch nicht gegeben hatte. Der berühmte Ingenieur und „Sicherheitspapst“ Béla Barényi – Portrait ab S. 200 – setzte hier erstmals seine Visionen ausgereift in einer neuartigen Sicherheitskarosserie um. Knautschzonen an Front und Heck nahmen bei einem Unfall Aufprallenergie auf und schützten dadurch die Insassen. Ein Meilenstein der Automobilgeschichte, das den Erfolg der Stuttgarter Weltmarke weiter ankurbelte und dessen Prinzip bis heute in nahezu jedem Automobil zu finden ist. 1963 gelang Mercedes-Benz mit dem „600er“ (W 100) ein weiterer großer Wurf. Vor allem die Pullman-Variante ließen sich Regierungschefs gerne als Staatslimousine liefern, auch gab es Sondermodelle, wie das für Papst Paul VI. gebaute Landaulet. Zudem war der "Große Mercedes“ repräsentativer Begleiter vieler berühmter Persönlichkeiten wie Coco Chanel oder Elvis Presley – Geschichte ab S. 60 – sie alle entschieden sich für dieses sehr luxuriöse und mit 205 km/h Spitzengeschwindigkeit für damalige Verhältnisse ausgesprochen schnelle Fahrzeug. 1968 stach dann der mächtige 300 SEL 6.3 (W 109) aus der Masse und erlangte als automobiles Status-Symbol des wirtschaftlichen Aufschwungs in Deutschland Ruhm und einen immensen Bekanntheitsgrad. Die Marke Mercedes-Benz war nicht nur ganz oben angekommen, sondern inzwischen für die ganze Automobilwelt Synonym für Individualität und Luxus auf besonderem Niveau, eben, Sonderklasse eben.

1972 erhielten die Oberklasselimousinen, mit denen Mercedes-Benz seit Jahrzehnten automobile Standards setzt, schließlich einen eigenen Namen. Auf dem Autosalon in Paris erblickte die neue S-Klasse der Baureihe 116 das Licht der Welt: Ein ganz besonderer Ort für Mercedes-Benz, an dem 1902 der allererste Mercedes debütierte.
So ist nun auch 1972 die Weltpresse sehr gespannt, die Belegschaft der Daimler-Benz AG in Stuttgart voller Erwartung und Stolz, denn mit der neuen Baureihe, verlässt erstmals ein Automobil die Werkshallen, das schon vor der Premiere vollste Vollendung verspricht. Und das war sie auch, die erste offizielle S-Klasse. Perfektion in Technologie, Verarbeitung und Luxus - alles in Serie und „Made in Germany“ noch dazu. Eine Besonderheit in der bisher existierenden automobilen Oberklasse. Eine, die es bis dahin so noch nicht gab.

In der Tat wurde deutlich, dass das „S“ der S-Klasse für etwas Besonderes steht, für eine Sonderklasse, obwohl man es durchaus beliebig mit anderen hochwertigen Superla-

At the same time, they also represented a new type of automobile, incorporating previously unheard-of safety standards into a production vehicle. With this series, the renowned engineer and “safety guru” Béla Barényi (portrait on page 200) was able to implement his vision for a completely new sort of car body. In the event of a collision, crumple zones in the front and rear would absorb the force of impact, protecting those inside. This milestone in automotive history accelerated the rise of Stuttgart’s world brand, and this safety principle is now integrated into nearly every automobile.
Mercedes-Benz clocked up yet another success with the new 600 (W 100) introduced in 1963. Countless heads of government ordered its Pullman version as a state limousine, and there were also countless special versions such as the Landaulet built for Pope Paul VI. This “great Mercedes” was also seen in the company of many prominent personalities such as Coco Chanel and Elvis Presley (story on page 60). They were all drawn to this opulent vehicle with a top speed of 205 km/h; a particularly fast vehicle for the time.
In 1968, the powerful 300 SEL 6.3 (W 109) raced ahead of the rest of the field, achieving great fame and renown as the automotive status symbol of Germany’s economic miracle. The Mercedes-Benz brand had done more than just arrive at the top of the automobile world. It had become a synonym for individuality and luxury at the highest level; a truly special class.

It was in 1972 that the Mercedes-Benz range of premium automobiles – vehicles with which the firm had been setting standards for decades - finally received its own name. At the Paris Auto Show, the new S-Class series 116 was unveiled to the world. Paris was already a special place for Mercedes-Benz because it was there that the very first Mercedes had made its debut in 1902. And once again in 1972, the world press watched with great interest, and every employee of Daimler-Benz AG looked on with pride. For the first time, even before its official premiere, as the vehicle began rolling off the assembly line, it was rumoured to have attained a new level of automotive perfection, down to the smallest detail. And so it was: This was the first official S-Class. A new vehicle integrating perfection in technology, workmanship and luxury, with the added allure of “Made in Germany”, which was a new and unique differentiator in the premium class of the time.

It quickly became clear that the “S” in the S-Class did in fact stand for something very special indeed. A true superlative – quite simply, a sensation. Because the 116 series didn’t just introduce a variety of new production-ready technologies, it actually made them the topic of worldwide discussion. For example, it introduced new safety features such as the revolutionary antilock braking system (ABS); today an automotive standard and in many countries now a legal requirement. Since this point, the S-Class has remained a trendsetting forerunner for the entire automotive world – an engineering masterpiece that has rarely been surpassed in safety, luxury, quality and

Mercedes-Benz 450 SE der Baureihe W116 in 3 charakteristischen Farben der 70er Jahre mit den charakterisierenden waagrecht orientierten großflächigen Scheinwerfern.
Mercedes-Benz 450 SE of the W116 series in three colours typical of the 1970s, with its characteristic wide rectangular headlights.

tiven austauschen könnte: Superklasse, Spitzenklasse. Aber man könnte auch einfach Sensation sagen. Denn die Baureihe 116 brachte nicht nur neue, serienreife Technologien hervor, sondern diese auch weltweit ins Gespräch. Zum Beispiel Sicherheitsstandards, wie das revolutionäre Antiblockiersystem, kurz ABS, heute quasi automobiler Standard und in vielen Ländern sogar Pflicht. Seitdem ist die S-Klasse der richtungsweisende Technologieträger der Autowelt. Stets aufs Neue. An Sicherheit, Luxus, Qualität und Individualität kaum zu überbieten. Ihren ersten Zenit hatte sie 1979, als die Baureihe 126 den „116er" ablöste. Sie wurde zum erfolgreichsten Oberklassen Mercedes-Benz aller Zeiten

Königin der Niederlande. Die S-Klasse war spätestens von da an auch der breiten Masse als Innovationsträger und Repräsentationsfahrzeug mit Weltstatus bekannt.
Das „S" bekam schließlich im Jahr 1993 eine stärkere Gewichtung, als der bisher nachgestellte Buchstabe durch eine Nomenklaturänderung vor die dreistellige Zahl gesetzt wurde, als wegweisendes Symbol für den technologischen Fortschritt.

Wer in der aktuellen Baureihe 222 Platz nimmt und eine Runde mit ihr fährt, angetrieben etwa durch den neuen Hybrid-Motor mit Plugin-Technologie oder die BlueTec-Aggregate der neuesten Generation, im Inneren den Ma-

FÜR DAS BESTE „MADE IN GERMANY" STEHT BEI MERCEDES-BENZ AB 1972 EIN KÜRZEL: „S"

- Weltmarkführer und innovativer Technologieträger. Ausgerüstet mit weiteren bahnbrechenden Meilensteinen der Technik, wie dem ersten Airbag kombiniert mit einem pyrotechnischen Gurtstraffer. An Zuverlässigkeit und damit auch an Beliebtheit kaum zu toppen, wurde er in verschiedensten Ausführungen bestellt, sogar Sondermodelle wurden aufgelegt wie Staatslimousinen oder sogar ein S-Klasse-Modell für den Vatikan. Und er war beliebt als Basis für Spezialanfertigungen, die über Zweitfirmen realisiert wurden, wie ein S-Klasse-Cabriolet für Juliana von Oranien-Nassau, die damalige

terialmix auf sich wirken lässt und den Platz genießt, weiß genau, das hier ist die Referenz der automobilen Oberklasse, das ist der Leitstern. Und wer dann die Geschichte der S-Klasse Revue passieren lässt und über ihre Wurzeln nachdenkt, sich dazu vorstellt, dass dieses Automobil schon völlig autonom fahren könnte, der wird die S-Klasse schließlich ganz verstehen. Ihre Sprache ist Faszination, ihre Botschaft eine klare Ansage, die exakt das auf den Punkt bringt, was sie schon immer verkörpert hat: Führungsanspruch. •

individuality. The S-Class reached its first zenith in 1979, when the 126 series succeeded the 116, quickly becoming the most successful premium-class Mercedes-Benz of all time, the world market leader, and an innovative trendsetter. The vehicle was equipped with

featuring plug in technology or the latest BlueTec aggregate engine. They will slowly take in the mix of fine materials and savour the generous space, and know with certainty that they are looking at the lodestar for the entire premium class. Likewise, anyone who

FROM 1972, THE "S" STANDS FOR THE BEST OF GERMAN CRAFTSMANSHIP

further groundbreaking technological milestones such as the first airbag, combined with a pyrotechnically-powered seatbelt pre-tensioner. Supremely dependable and thus highly popular, this S-Class was delivered in a wide range of body styles, including official state limousines and a special model for the Vatican. It also became popular as the basis for custom vehicles produced by other firms, such as an S-Class convertible for Juliana of Orange-Nassau, then Queen of the Netherlands. The S-Class had truly established itself in the eyes of the wider public as an innovator and worldwide status symbol.
The year 1993 brought a heavier weight for the "S" designation. Previously placed after the three digit model number, the Mercedes-Benz nomenclature was revised to place the S before the number, as a prominent symbol of technological innovation.

Anyone fortunate enough to take a seat in the current 222 series and take it for a spin will be powered by either the new hybrid motor

takes a short excursion into the history of the S-Class to consider its roots and heritage, who is able to imagine that this automobile might one day drive autonomously, will finally begin to understand the S-Class at its very core. Its language is fascination. Its message is an unmistakable proclamation, one that summarises everything that the car has ever embodied in one singular concept: the claim to leadership. •

Mercedes-Simplex 60 PS Reiselimousine von Emil Jellinek.
Emil Jellinek's Mercedes-Simplex 60 hp touring saloon.

SIMPLEX

1903

Der Mercedes-Benz 300 (W 186) ist die Sensation der Internationalen Automobilausstellung IAA im April 1951 in Frankfurt/M. und mit 160 km/h Spitze auch das schnellste deutsche Serienautomobil. Mercedes-Benz etabliert sich damit – nur sechs Jahre nach dem Zweiten Weltkrieg – wieder als Hersteller von Luxus-Automobilen.

The Mercedes-Benz 300 (W 186) was the sensation of the Frankfurt Motor Show in April 1951. With a top speed of 160 km/h, it was also the fastest German car. The W 186 helped Mercedes-Benz re-establish itself as a luxury automobile manufacturer just six years after the end of the Second World War.

W 186

19
51

300 SE
D
S-KV 559

Der „Heckflossen-Mercedes“ besaß als erstes Serienautomobil überhaupt eine stabile Fahrgastzelle und wirksame Knautschzonen.
The Mercedes “Fintail” was the first production car with a stable passenger cell and functioning crumple zones.

W 111/112

1959

1978 feiert das elektronisch gesteuerte Anti-Blockier-System ABS seine Serienpremiere in der S-Klasse der Baureihe 116.
The electronic antilock braking system (ABS) made its debut in a production car in the 116 series, an S-Class series, in 1978.

W/V 116

1972

1981 hält der Airbag, ein zentrales Element der Automobilsicherheit, in der S-Klasse Einzug in den Serienautomobilbau.
In 1981, the first series-produced car to include an airbag – a central element in automotive safety – was a Mercedes-Benz S-Class.

W/V 126

1979 –

1991

Von 1991 bis 1998 wurden über 400.000 S-Klassen der Baureihe 140 hergestellt.
Over 400,000 S-Class automobiles in the 140 series were manufactured between 1991 and 1998.

W/V 140

Bei der Baureihe 220 wurden Gewicht und Luftwiderstand gesenkt, was bessere Fahrleistungen und Verbrauchswerte zur Folge hatte.
In the 220 series, weight and air resistance were lowered, improving driving performance and fuel consumption.

W/V 220

19₉₈

Von 2005 bis 2013 war die Baureihe 221 weltweit durchgehend Marktführer im Segment der Oberklasse-Limousinen.
The 221 series was the global market leader in the grand saloon segment from 2005 to 2013.

W/V 221

2005

S 350
S MB 1091

Im Mai 2013 erscheint die neue S-Klasse Baureihe 222 mit zahlreichen Innovationen.
The new S-Class 222 series, featuring a range of innovations, premiered in May 2013.

W/V 222

2013

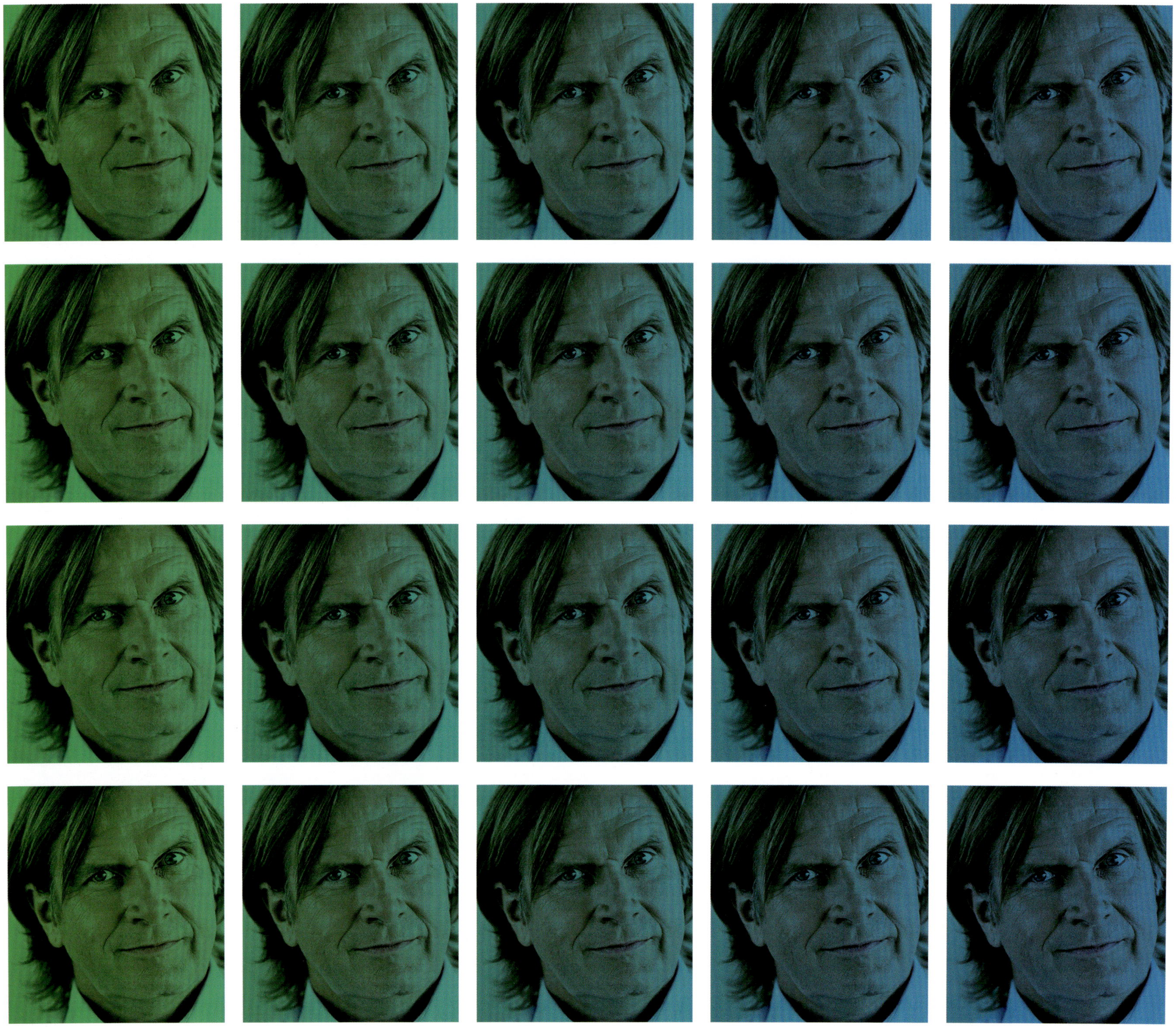

„DIE FORM IST TEIL DER BOTSCHAFT“

“THE APPEARANCE IS PART OF THE MESSAGE”

BEI DER MERCEDES-BENZ S-KLASSE ALS GESAMTKUNSTWERK SPIELT AUCH DIE GESTALTUNG EINE ZENTRALE ROLLE. WIE DAS GENAU FUNKTIONIERT? FRAGEN WIR DOCH MAL HANS-DIETER FUTSCHIK, DIRECTOR EXTERIOR DESIGN PASSENGER CARS BEI MERCEDES-BENZ. ER IST FÜR DAS AKTUELLE DESIGN DER S-KLASSE VERANTWORTLICH.
EIN INTERVIEW VON MICHAEL KÖCKRITZ.

FOR THE MERCEDES-BENZ S-CLASS, A “GESAMTKUNSTWERK”, DESIGN IS ONE OF THE KEY ISSUES. WHAT EXACTLY DOES THIS ENTAIL? HANS-DIETER FUTSCHIK, DIRECTOR OF EXTERIOR DESIGN FOR PASSENGER CARS AT MERCEDES-BENZ, HAS AN ANSWER. HE IS RESPONSIBLE FOR THE DESIGN OF THE CURRENT S-CLASS. AN INTERVIEW WITH MICHAEL KÖCKRITZ.

S KR 1142
S KL 4334
S KT 4998

„Die S-Klasse steht für alles, was uns wichtig ist."
The S-Class represents everything that matters to us.

Die S-Klasse steht für das Beste vom Besten. Wie drückt sich das im Design aus?
Das Design eines Autos prägt stets den ersten Eindruck. Also signalisiert die S-Klasse so einen Anspruch natürlich auch mit den Mitteln der Gestaltung. Die Form ist immer der erste Teil der Botschaft eines Autos. Innerhalb unserer Modellreihen ist das Design der S-Klasse immer das ausbalancierteste und ausgereifteste, am Ende ist alles sehr präzise ausbalanciert und ideal auf den Punkt gebracht. Nur so erzielt man die souveräne Langzeitwirkung, die dieses Modell authentisch prägt. Das Design der S-Klasse ist ein Statement wie das ganze Auto selbst. Eine S-Klasse zu gestalten ist immer etwas ganz besonderes.

Ist diese Langzeitwirkung ein Merkmal von Premium-Design? Also jenseits von Moden?
Die S-Klasse wirkt jenseits von schnelllebigen Moden, obwohl sie immer ein klares Statement in ihrer jeweiligen Zeit ist. Die S-Klasse kennt und beherrscht das Spiel mit zeitgenössischen Gestaltungsentwicklungen, aber sie ist nie ein modisches Auto. Das ist ein essentielles Merkmal des Autos.

Jetzt steht die S-Klasse im Markenanspruch auf der Höhe der Zeit. Wie wichtig ist Design für diese Positionierung der Marke?
Die S-Klasse ist der Mercedes unter den Mercedessen, sozusagen das Kernauto der Marke und steht für alles, was uns wichtig ist und was wir eigentlich so tun.
Dieses Auto steht für unseren Ursprung, da kommen wir her. Wir stehen für Luxus- und Qualitätsfahrzeuge. Die S-Klasse ist das Herz der Marke. Und genau das darf man auch erleben, wenn die S-Klasse vor einem steht.

Welche Merkmale des Designs prägen die S-Klasse? Welche Linien, welche Proportionen?
Zunächst wirkt natürlich die elegante Silhouette als Ganzes, dann das Volumen in Verbindung mit dieser Silhouette und einer sehr klaren Linienführung. Letztlich zählt aber bei diesem Auto vor allem diese perfekte Balance aus Dynamik und Ruhe. Das Auto vermittelt Dynamik und ruht dabei doch immer in sich selbst. Wahrscheinlich ist es gerade die Auflösung dieses scheinbaren Widerspruchs, der die S-Klasse optisch so nachhaltig attraktiv macht. Fronterscheinung setzt auf ein starkes, unverwechselbares, charismatisches Gesicht. Die S-Klasse hat diesen typischen Mercedes-Blick: Selbstbewusst und stolz, entschlossen und über den Dingen stehend, dabei nicht im Geringsten grob und klotzig und nie einschüchternd. Das Gesicht der S-Klasse ist ebenso sympathisch wie souverän dynamisch und selbstbewusst. Hier passt eben alles sehr harmonisch zusammen.

Das war außen. Wie ist es denn innen? Woran erkennt man, wenn man in dem Wagen sitzt, über das Interieur den Designanspruch?
Die S-Klasse ist das Auto, bei dem wir alle unsere Karten ziehen, die man im Kontext von Luxus nur sinnvoll spielen kann. Also umarmen wir die Insassen geradezu mit allem, was die sich nur wünschen können, wenn es um ein intensives Luxusempfinden geht. Das Auto wird zum sinnlich-perfekten Luxus-Kokon. So etwas beginnt bei einer Linienführung, die die Insassen geradezu umarmt.
Die Linien der Instrumententafel hören an deren Funktionsgrenzen nicht einfach auf, sondern setzen sich fließend in der Gestaltung der Türen harmonisch fort und von dort im ganzen Innenraum. Alles ist stimmig umhüllend

The S-Class represents the absolute cream of the crop. How is this excellence expressed in its design?
The way a car is designed always affects the first impression it makes. And the S-Class establishes these standards by means of its design as well. Its lines are always the first part of the message a car sends. The design of the S-Class is always the most balanced and mature among all of the different kinds of Mercedes models: ultimately everything has been very precisely equilibrated and brought to the perfect conclusion. This is the only way to create the long-term effect of superiority which is the authentic mark of this model. The design of the S-Class is as much of a statement as the car itself is. Designing an S-Class is always something very special.

Is this long-term effect a hallmark of premium design? Being above and beyond trends?
The S-Class has a classic air to it – it is far removed from flash-in-the-pan fashions, even though it always makes a clear statement within its own time. The S-Class understands the game with contemporary developments in design and is a master of them; it has style but is never trendy. That is one of its essential characteristics.

In terms of the brand's standards, the S-Class has its finger on the pulse. How important is design when it comes to positioning this brand?
The S-Class is the ultimate member of the Mercedes family. It can be seen as the core of the Mercedes brand, and it represents everything which matters to us and everything we do. This car reflects its origins: it shows where we come from. We stand for luxury and quality cars. The S-Class is the heart of the brand, and that is exactly what you get to experience when you have one standing before you.

What elements of design define the S-Class? What lines and proportions?
At first, of course, the elegant silhouette has an effect all on its own, and then you take in the other contours in conjunction with this silhouette and extremely clear lines. What really distinguishes this car, however, is the perfect balance it strikes between dynamic energy and tranquility. The car radiates energy but still gives off a sense of peace within itself. The fact that the S-Class manages to resolve this apparent contradiction is probably what makes its aesthetic so enduringly appealing. The car's appearance from the front emphasizes a strong, unmistakable, charismatic face. The S-Class has that typical Mercedes look: self-confident and proud, decisive and in control, but in no way coarse or clunky, and never intimidating. The face of the S-Class is in equal parts sympathetic, independently dynamic and self-assured. Everything comes together harmoniously.

That's the exterior. What does it look like inside? When you sit down in the car, what design standards will you see on the interior?
The S-Class is the car where we pull out every one of the stops to offer everything which makes sense in the context of automotive luxury. This means we all but embrace the passengers with everything they could possibly want in terms of experiencing the utmost in opulence. The car becomes a perfect cocoon of physical magnificence. And that starts with the lines, which practically wrap their arms around you. The lines of the instrument panel do not simply end where their functional limits are; they continue

Handwerkskunst: Eine Design-Skizze von Robert Lesnik zur neuen S-Klasse.
A hand-drawn work of art: a design sketch by Robert Lesnik for the new S-Class.

durchgestaltet. Ein weiterer wichtiger Aspekt ist die hohe Qualität der Materialien. Für uns lebt Luxus-Interieur-Design klassischer Art immer von einem konsequent wertigen Dreiklang aus Leder, Holz und Metall. Und dann zählt natürlich die handwerkliche Qualität. Dieses Können macht ein Luxuserlebnis letztlich wesentlich aus. Die S-Klasse steht für diese Präzision im Detail. Also bemühen wir uns wirklich um jedes kleinste Detail, vom Sicherungsstift in der Tür über die Luftdüsen bis zu den Hebeln am Lenkrad. Ich denke, auch bei der aktuellen S-Klasse ist uns das wieder richtig gut gelungen.

Gibt es denn nicht unterschiedliche Luxus-Anforderungen unter den Kunden der S-Klasse?
Doch, natürlich. Und wir wollen sie gerne alle bedienen. Es gibt Kunden, die machen Luxus vor allem an einer High-Tech-Modernität fest, andere mögen es wiederum in erster Linie superkomfortabel, hin und wieder gerne auch etwas neo-barock. Die Kunst ist es, eine ideale Schnittfläche zu entwickeln, die von allen als ihre echte S-Klasse erlebt werden kann.

Welche Bedeutung hat die Farbigkeit? Also die Farben, die gespielt werden?
Innen oder außen?

Beides.
Farbe ist ein wesentliches Element der Individualisierung. Erstaunlicherweise entscheiden sich dennoch die meisten unserer Kunden immer nur für zwei Farben. Silber und Schwarz sind hier nach wie vor die beliebtesten Hauptfarben. Im Angebot haben wir selbstverständlich alles, was sich unsere Kunden wünschen könnten. S-Klasse-Fahrer sollen ja ihr ganz persönliches Auto konfigurieren können, da spielt Farbe eine wichtige Rolle.

Würden Sie einen Ausblick auf die Design-Entwicklung der S-Klasse verraten, wie sich das Design entwickeln wird? Wird der Wagen größer, wird er flacher?
Natürlich stellen wir uns jedes Mal am Anfang von einer S-Klasse-Entwicklung die grundsätzliche Frage, ob dieses Konzept überhaupt grundsätzlich noch zeitgemäß ist. Und jedes Mal erkennen wir, dass es aktuell und attraktiv wie eh und je ist. Die S-Klasse als eine wegweisende, große, repräsentative, klassische Luxus-Limousine, das wird nach wie vor bleiben. Und jedes Mal kommen wir auch zu dem Schluss, dass das Designkonzept nach wie vor schlüssig ist und dieselben Aufgaben zu erfüllen hat wie von Anbeginn an.

Gibt es einen Lieblingsblick von Ihnen auf das Auto? So eine Lieblingsperspektive?
Üblicherweise sehe ich mir die Autos immer gerne von schräg vorne an, so ein bisschen aus der extremen Perspektive, sodass man das Gefühl hat, das Auto kommt einem entgegen und man kann bereits die Dynamik der seitlichen Linienführung erkennen. Hier spielt ja bereits alles wunderbar zusammen. Der Blick auf die Front und wie sie so in die Seite übergeht, ist für mich immer einer der attraktivsten. Aber wenn man so lange an einem Auto gearbeitet hat, ist einem schließlich jede Stelle irgendwie ans Herz gewachsen halt. Da hat jedes Detail seine Ästhetik und Wirkung. •

Vielen Dank für das Gespräch.

INTERVIEW MICHAEL KÖCKRITZ

to flow harmoniously into the design of the doors, and from there they extend throughout the interior. Everything has been designed to come together as a sumptuous whole that surrounds you. Another important aspect is working with top-tier materials. In our eyes, premium leather, wood and metal are the building blocks which form the cornerstone of designing classic luxurious interiors. And then of course, quality craftsmanship matters: the expertise of our work ultimately defines a true luxury experience. The S-Class expresses this precise attention to detail. So we put great effort into refining even the smallest detail, from the lock pin in the door to the air vents to the hand gears on the steering wheel. I think we have effectively managed the task once again in the current S-Class.

Aren't there different luxury standards among the S-Class clientele?
Of course there are, and we want to satisfy all of them. We have clients who see luxury as being a question of high-tech innovation first and foremost, whereas others want more than anything for their cars to be incredibly comfortable. Every now and then there are people who tend towards something slightly neo-baroque. The true art is developing the perfect canvas, as it were, one which means everyone can experience their own true S-Class.

What does the colour scheme represent – the colours being emphasized?
Inside or outside?

Both.
Colour is a key element when it comes to customizing a car. Surprisingly, though, most of our clients select only one of two colours: silver and black are the perennial favourites. It goes without saying that our palette of choices includes anything our clients could want. People who drive an S-Class should be able to configure their car exactly to their liking, and colour is an important aspect of that.

Could you give us a hint about the designs you're developing for the S-Class? Which way is it headed? Will it be larger, or perhaps lower?
Obviously every time we start to develop a new S-Class, we ask ourselves the fundamental question as to whether or not the overall concept is still contemporary. And every time we realize it is as up-to-date and attractive as it ever was. The S-Class as a trail-blazing, large, prestigious, classic luxury automobile – that will remain unchanged. And every time, we come to the conclusion that the design concept is still solid, and it still has to meet the same criteria as it has from the beginning.

Do you have a favourite angle for looking at a car? A favourite point of view?
Usually I like to have a look at a car from a diagonal front angle – get a glimpse of it from an extreme perspective as if it were driving towards you. This lets you see the dynamic lines of the sides. Everything comes together beautifully from that angle. Having a look at the front as it transitions into the side is one of the best views, in my eyes. That said, if you spend long enough working on a car, at some point you grow to love it from every angle. Every detail has its own aesthetic and effect. •

Thank you for the interview.

INTERVIEW MICHAEL KÖCKRITZ

1 S-KLASSE Die Form ist Teil der Botschaft **S-CLASS** The appearance is part of the message

S-SY 946

50 JAHRE YEARS

DER MERCEDES-BENZ 600 (W 100) VON 1963 IST DAS MUSTERBEISPIEL FÜR DAS, WAS PASSIERT, WENN MAN DIE S-KLASSE MIT NOCH ETWAS MEHR LUXUS MULTIPLIZIERT. EIN AUTOMOBIL FÜR DIE HELDEN DER DAMALIGEN ZEIT.

THE 1963 600 MERCEDES-BENZ (W 100) IS THE ULTIMATE EXAMPLE OF WHAT HAPPENS WHEN YOU MULTIPLY THE S-CLASS BY EVEN MORE LUXURY. A CAR FOR THE HEROS BACK IN THE DAY.

MERCEDES-BENZ 600

Es gibt Autos, die von ihrer Zeit geprägt wurden. Und es gibt einige wenige Automobile, die ihre Zeit geprägt haben. Der Mercedes-Benz 600 ist so eine Ausnahmeerscheinung. Kaum ein anderes Fahrzeug weltweit hat bis heute einen solchen Wiedererkennungseffekt wie das intern schlicht W 100 genannte Fahrzeug. Pardon – Fahrzeug? Es stimmt, auch der „600er" hat vier Räder und ein Lenkrad und einen Motor. Doch wenn je der Begriff Fahrzeug allein nicht ausreichte, dann bei diesem Automobil. Wo der „600er" in den Sechzigern und Siebzigern des letzten Jahrhunderts auftauchte, wichen die Menschen zur Seite. Nicht aus Furcht, sondern aus Ehrfurcht. Denn auf der einladenden Rücksitzbank saßen die Stars der damaligen Zeit. Liz Taylor ließ sich in ihm chauffieren, John Lennon, auch Elvis Presley. Roger Moore? Der wurde als James Bond in „Octopussy" von ihm verfolgt. Und hat daraufhin sicher von ihm geträumt. Sicher ist, die Stars schätzten und liebten das Auto, und ihr Publikum sah und liebte das Auto und dessen unwiderstehliche Strahlkraft. Der „Große Mercedes", mit über 200 km/h damals eines der schnellsten Serienfahrzeuge der Welt, polarisierte. Und es provozierte, allein durch seinen opulenten Auftritt – und stimmte im selben Moment fast wieder versöhnlich. Jackie Kennedys zweiter Ehemann, der Großreeder Aristoteles Onassis genoss das Flair ebenso wie Jack Nicholson. Wenn aber durch seine steil stehenden Seitenscheiben weniger beliebte Zeitgenossen wie Irans Schah Reza Pahlavi oder der sowjetische Führer Leonid Breschnew den wartenden Menschenmassen huldvoll zuwinkten und die Standarten auf den vorderen Kotflügeln majestätisch im Wind flatterten, wussten die Zaungäste dennoch zwischen Ross und Reiter zu unterscheiden. Drinnen Diktator, draußen Renommée. Und sie wussten: Dieses Auto krönt nicht den Insassen. Es krönte den Automobilbau.

Mit seiner stattlichen Länge von 5,54 Meter (6,24 Meter in der berühmten Pullman-Variante mit verlängertem Radstand) hatte es durch aus beste Voraussetzungen, als protzig und dekadent tituliert zu werden. Nicht so der „600er". Der 1963 auf der IAA in Frankfurt vorgestellte Wagen war vom ersten Zeichenstrich an als Staatskarosse, als Repräsentations-Limousine konzipiert gewesen. Wer in ihm Platz nahm und in der mit ausgesuchten Edelhölzern und Ledern ausgeschlagenen Karosse die Beine ausstreckte, war wer oder hielt sich zumindest dafür. Und er oder sie waren in besserer oder bester Gesellschaft. Papst Paul VI. diente der „Große Mercedes" in einer Landaulet-Fassung zwanzig Jahre als irdisches Gefährt – bevor es 1985 nach Deutschland zurückkehrte und seitdem im Mercedes-Benz Museum die staunenden Gäste bis heute in seinen Bann zieht. Wer einmal um dieses Automobil herum-

There are cars that are defined by their time, and there are even a few cars which define their time. The Mercedes-Benz 600 is one of these precious few. Hardly any other vehicle anywhere in the world is as recognisable as the car known in-house simply as the W 100. I'm sorry – did you say "car"? It's technically true; the 600 has four wheels, a steering wheel and an engine. But if ever there were a case when referring to something as a car does not do it justice, then it is here. In the '60s and '70s, when a 600 passed by, people stepped aside and took notice. Not out of fear, but out of awe, since the roomy backseat was where the heros of the day sat. Liz Taylor was chauffeured around in one, and so were John Lennon and Elvis Presley. Roger Moore was chased in a 600 as James Bond in "Octopussy" and certainly dreamed of one after that. Celebrities openly loved and appreciated the vehicle, and their audiences also saw and loved it and its irresistible allure. Reaching top speeds of over 200 km/h, the "Big Mercedes" was one of the world's fastest series models, and it was polarising. Its sumptuous aesthetic was downright provocative and yet seemed almost comforting at the same time. Jackie Kennedy's second husband, shipping magnate Aristotle Onassis, enjoyed its flair just as much as Jack Nicholson did. On some occasions, however, as the standards in the front fenders majestically fluttered in the wind, the steeply angled side windows were lowered to reveal that the person inside waving benevolently to the waiting masses was someone like the Iranian Shah Reza Pahlavi or Soviet leader Leonid Brezhnev. But even at such moments, the people on the sidelines could distinguish between the horse and its rider, as it were. There might have been a dictator on the inside, but the outside was still pure prestige. And people understood – this car was not simply the crowning glory of its passengers; it was the crowning glory of automotive manufacturing itself.

At a majestic length of 5.54 metres (or 6.24 in the famous Pullman limousine with a long wheel base), it was the perfect car to be declared pretentious and decadent. But the 600 was destined for more. First introduced at the IAA in Frankfurt in 1963, it was designed from the outset as an official car, a state limousine. If you had the privilege of stepping in and stretching your legs amidst the fine woods and premium leather of the interior, you were somebody, or at least you thought you were. And being in that car meant you were in fine company, if not the finest. A landaulet "Big Mercedes" served as Pope Paul VI's earthly conveyance for twenty years before it returned to Germany in 1985, and since then it has captivated astounded visitors at the Mercedes-Benz Museum. If you walk around this car to look at it from every angle, you will end up covering a distance of around sixteen metres

wandert, hat dann am Ausgangspunkt gut sechzehn Meter Fußweg bewältigt – und nebenbei ein bedeutendes Stück Automobilgeschichte erlebt.

Die Luxuslimousine namens Mercedes-Benz 600 prägte nicht nur ihre Zeit, sie war ihr auch voraus. Sie setzte nicht nur in Größe und Gewicht (zwischen 2,5 und 3,3 Tonnen je nach Ausführung) Maßstäbe. Ein revolutionäres Hydrauliksystem erfasste alle beweglichen Teile: Lautlos glitten die Scheiben hinunter oder neigte sich die Rücklehne in eine kommode Position. Servolenkung und -bremse waren ebenfalls Serie, damals aber alles andere als selbstverständlich. Ein V8-Einspritzmotor mit 6,3 Liter Hubraum war eine komplette Neuentwicklung an der Spitze des vor einem halben Jahrhundert technisch Machbaren und bescherte dem W 100 gar den Titel des „größten Sportwagens aller Zeiten". Davon ließ sich wohl auch Schlagersängerin Manuela beeindrucken, als sie etwas abgewandelt „Schuld war nur der Bossa Nova" sang und damit 1963, dem Geburtsjahr des „600ers", schlagartig berühmt wurde. Die junge Berlinerin besaß einen roten „600er" und fuhr damit auf die Bühne ihrer Shows im Dunes Hotel in Las Vegas.

Und während sich neben vielen anderen Persönlichkeiten Herbert von Karajan, Japans Kaiser Hirohito, Modedesignerin Coco Chanel oder Vorzeige-Playboy Gunter Sachs im „600er" chauffieren ließen und Ivan Rebroff oder auch Udo Jürgens es ihnen gleich taten, machte der W 100 auch im Filmgeschäft Karriere. In „Im Geheimdienst Ihrer Majestät" von 1969 stirbt die einzige filmische Ehefrau von James Bond durch einen Schuss ausgerechnet aus dem Fond eines „600ers". Auch in den James-Bond-Filmen „Diamantenfieber" und „Octopussy" hatte er unvergessene Auftritte.

Der unübertreffliche und bis heute unübertroffene Mercedes-Benz 600 stellt – in aller gebotenen Bescheidenheit – seine direkten Nachfahren bis heute in den Schatten. Als eine der größten automobilen Persönlichkeiten hat er Kanzler, Kirchenoberhäupter und Präsidenten in sich aufgenommen. 1981, nach achtzehn Jahren Bauzeit und ohne das, was wir heute Facelift nennen, trat er würdevoll ab. Damals nicht und bis heute nicht hat je ein deutsches Konkurrenzprodukt gewagt, den Nimbus des großen, unantastbaren Mercedes-Benz anzutasten. Auch wenn der Alte heute nur noch stumm im Museum steht, umfasst jeden Betrachter ein Hauch von Würde, Korrektheit und Etikette. Unbewusst prüft so mancher den korrekten Sitz von Hemd und Krawatte.

Ein Auto wie ein Denkmal? Nein. Eher ein Denkmal von einem Automobil. •

– and in the process, you will experience a significant part of automotive history.

The luxury saloon known as the Mercedes-Benz 600 not only defined its day, it was ahead of its time. It set benchmarks, and not simply because of its size and weight (between 2.5 and 3.3 tonnes, depending on the model). A revolutionary hydraulic system handled all of the moving parts: the windows opened and closed silently, and the seat reclined into a comfortable position. Its serial power steering and brakes were by no means to be taken for granted at the time. The 6.3 litre V8 injection engine was a total innovation at the very peak of what was technically feasible a half-century ago, and it even earned the W 100 the title "the world's largest sports car of all time." Amongst those it impressed was the German popular singer Manuela, as she skyrocketed to fame thanks to a version of "Blame it on the Bossa Nova" with lyrics glorifying 1963, the year when the 600 was "born". The young Berlin native had a red 600 and used to drive it on stage during her shows at the Dunes Hotel in Las Vegas.

Countless personalities became dedicated fans of the car, including conductor Herbert von Karajan, notorious playboy Gunter Sachs, Japanese emperor Hirohito, German-Russian singer Ivan Rebroff, fashion designer Coco Chanel and German recording artist Udo Jürgens, and at the same time the W 100 made a name for itself on the silver screen. In the 1969 "On Her Majesty's Secret Service", James Bond's only wife in the series was killed by a shot fired from the back seat of a 600. It also played unforgettable parts in the Bond movies "Diamonds Are Forever" and "Octopussy."

With all due modesty, the unsurpassable and still unsurpassed Mercedes-Benz 600 continues to leave its direct successors in the dust. As one of the greatest of automotive luminaries, it transported chancellors, church leaders and presidents. It was retired with dignity in 1981 after eighteen years of production and without ever undergoing what we call a face-lift. Neither then nor now has any German competitor attempted to aspire to the air of greatness surrounding this untouchable Mercedes-Benz. Even if it can now only be seen standing silently in museums, it still enshrouds everyone who sees it with a sense of dignity, decorum and mannerliness. The sight of it subconsciously prompts some people to adjust their shirt and tie. It is not so much a car like a monument – instead, it is more a monument of a car. •

Britischer Schauspieler 1927
Charmant. Markant. Direkt.
British actor, 1927
Charming. Striking. Direct.

ROGER MOORE

Als der britische Schauspieler, in „Leben und sterben lassen“ die Rolle des James Bond übernimmt und seinen Landsmann Sean Connery als damals wohl berühmtesten Leinwandhelden ablöst, ahnt er wohl noch nicht, dass er diese Rolle seines Lebens sieben Mal in zwölf Jahren spielen sollte (und es tatsächlich schaffte, das berühmte Martini-Zitat zu vermeiden). In „Octopussy“, der 1983 187,5 Millionen US-Dollar einspielte, fährt der undurchsichtige Antiquitätenhändler Kamal Khan einen Mercedes-Benz 600 mit langem Radstand. Überhaupt war ein schwarzer Mercedes-Benz in vielen Filmen das bevorzugte Fahrzeug der Bösewichte .

In taking over the role of James Bond in “Live and Let Die” from his fellow Brit Sean Connery, who was the most famous on-screen hero of his day, British actor Roger Moore could not have guessed that he‘d be playing the role of a lifetime, a total of seven times in the next twelve years (and actually managed to avoid the notorious martini quote). In the 1983 movie "Octopussy", which had a box-office take of $187.5 million, the shady antiques dealer Kamal Khan drove a Mercedes-Benz 600 with a long wheel base. The black Mercedes-Benz tended to be the favourite vehicle amongst many movie villains.

„WELL, THANK YOU, SIR. THAT'S A GREAT COMFORT.“
JAMES BOND IN OCTOPUSSY

“BIG GIRLS NEED BIG DIAMONDS.” LIZ TAYLOR

US-amerikanische-britische
Filmschauspielerin 1932-2011
Schön. Erfolgreich. Begehrt.
British-American movie actress,
1932-2011.
Beautiful. Successful. Desirable.

LIZ TAYLOR

Als “Katze auf dem heißen Blechdach” oder “Cleopatra” neben ihrem späteren zweimaligen Ehemann Richard Burton war sie eine der größten, wenn nicht die größte Diva Hollywoods. Sie wurde als eine der schönsten Frauen der Welt gefeiert, war eng mit Andy Warhol und Michael Jackson befreundet. Sie liebte es, ein Star zu sein, beklagte sich nicht wie berühmte Filmkollegen über den Verlust der Privatsphäre. Schauspielerei war nicht ihr Beruf, es war ihr Leben. Sie lebte großartig und großzügig, war acht Mal verheiratet und eine der reichsten Frauen der USA. *„Einige meiner besten Männer waren Hunde und Pferde“*. Das beste Auto der damaligen Zeit war kein Mann. Es durfte bleiben.

Appearing in "Cat On A Hot Tin Roof" and "Cleopatra", starring alongside her later twice-husband Richard Burton in the latter, she was one of the greatest, if not the greatest Hollywood diva of her era. She was celebrated as one of the world‘s greatest beauties and was close friends with Andy Warhol and Michael Jackson. She loved being a star, and unlike many famous colleagues, she did not complain about her privacy being invaded. Acting was not her job: it was her life. And she lived in grand and generous style, marrying eight times and becoming one of the richest women in the US. *“Some of my best leading men have been dogs and horses”*. The best car of her day was not a man, so there was no problem: it could stay.

Britischer Musiker und Komponist 1940-1980
Berühmt. Provokant. Genial.
British musician and composer, 1940-1980.
Famous. Provocative. Brilliant.

JOHN LENNON

Warum die Besten immer zu früh gehen müssen – wer versteht das schon? 1980 wurde der Mitbegründer der Beatles in New York ermordet. „Imagine“, einer seiner schönsten Titel der späteren Solokarriere, gilt für viele als der beste Popsong aller Zeiten. Lennon nahm sich und sein Genie richtig ernst. *„Wenn Affen Klavier spielen können, warum sollten Menschen nicht dazu singen?“* – Zitat John Lennon. Der Mann aus Liverpool, der sich vom Schüchternen zum Intellektuellen entwickelte, war überzeugt, dass Liebe allein der Schlüssel für eine bessere Welt ist. Und er liebte die Farbe weiß. Also war auch sein Mercedes-Benz 600 weiß.

Why do the good always have to die young? In 1980, one of the founding members of the Beatles was assassinated in New York. "Imagine", one of the loveliest songs of his later solo career, is regarded by many as the best pop song of all time. Lennon took himself and his genius seriously: the quote, *"If monkeys can play piano, why shouldn't people sing to it?"* is attributed to him. The man from Liverpool who blossomed from a shy lad to an intellectual was convinced that love alone is the key to a better world. And he loved the colour white, which is why his Mercedes-Benz 600 was white.

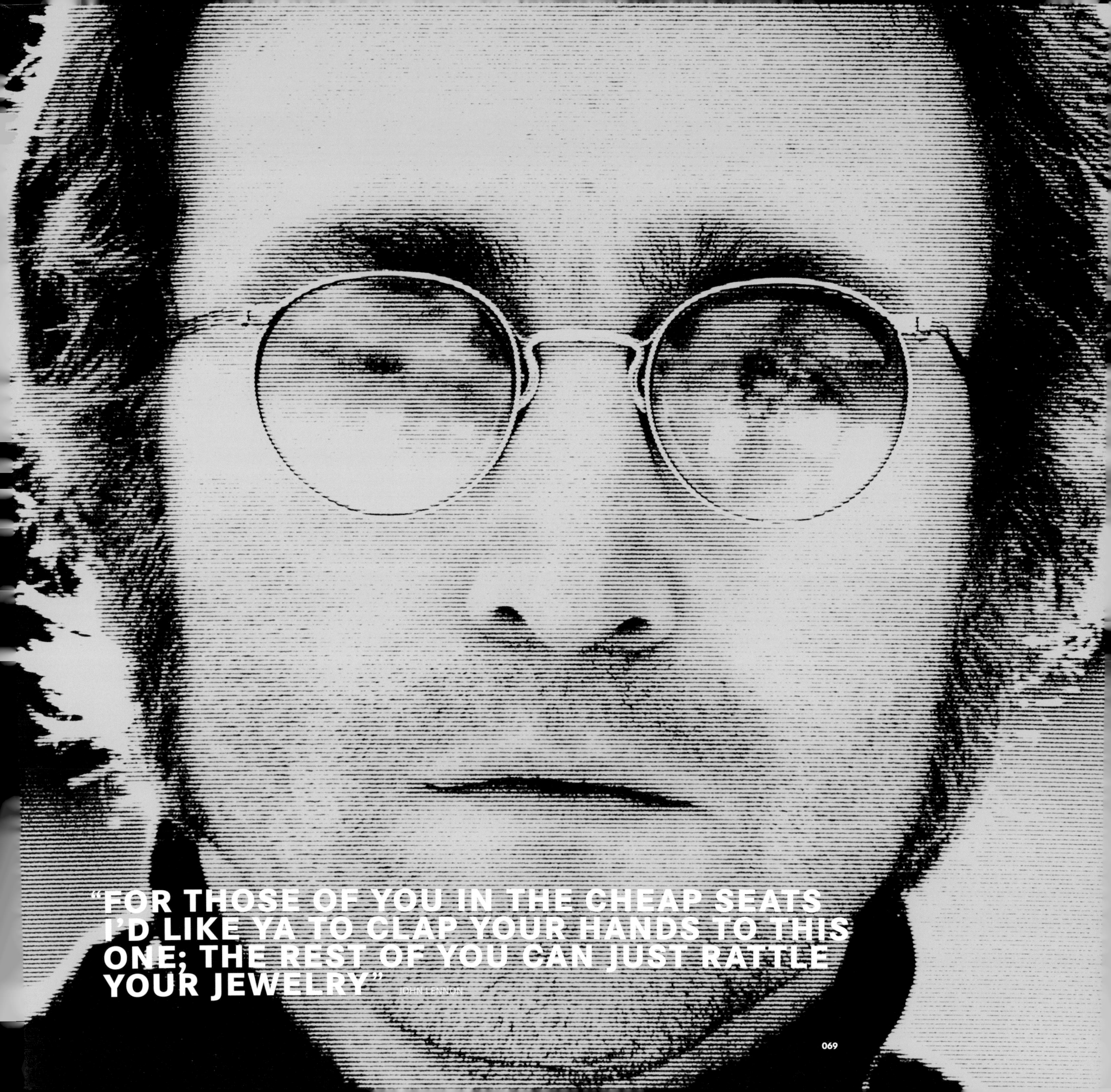

“FOR THOSE OF YOU IN THE CHEAP SEATS I’D LIKE YA TO CLAP YOUR HANDS TO THIS ONE; THE REST OF YOU CAN JUST RATTLE YOUR JEWELRY” JOHN LENNON

Französische Modeikone 1883-1971
Kreativ. Mondän. Stilvoll.
French fashion icon, 1883-1971
Creative. Sophisticated. Stylish.

“SOME PEOPLE THINK LUXURY IS THE OPPOSITE OF POVERTY. IT IS NOT. IT IS THE OPPOSITE OF VULGARITY.” COCO CHANEL

COCO
CHANEL

Die Gründerin des Modeimperiums Chanel ist eine der herausragendsten Persönlichkeiten der Mode-Geschichte. Sie machte nie Trends mit, sie setzte sie. Dabei begann die wohl stärkste Ikone der Haute Couture 1916 mit dem Nähen schlichter, aber funktioneller Damenmode: *„Ich habe die Frauen aus dem Korsett befreit."* 1920 folgten das berühmte „kleine Schwarze", 1922 ihr Parfum Chanel N°5. Die Freundin von Jean Cocteau, Igor Strawinsky, Pablo Picasso, Romy Schneider und Winston Churchill hatte schon 1936 über 4.000 Angestellte. Als sie 1971 im Alter von 87 Jahren in ihrer Pariser Suite diese Welt verlässt, soll sie gesagt haben: *„So stirbt man also."* Ihr Mercedes-Benz 600 überlebte seine berühmte Besitzerin und ging viel später in den Besitz von Popsänger Jamiroquai über. Was Coco wohl davon gehalten hätte?

The founder of the Chanel fashion empire is one of the most distinctive personalities in fashion history. She never followed trends: instead, she set them. The most powerful icon of haute couture got her start in 1916 by sewing simple but functional women's clothing, stating that she liberated women from their corsets. Her famous "little black dress" was introduced in 1920, and in 1922 she launched her perfume Chanel N°5. A friend of Jean Cocteau, Igor Stravinsky, Pablo Picasso, Romy Schneider and Winston Churchill, she had over 4,000 employees by 1936. In 1971, when she left this world at the age of 87, her final words in her Paris suite were said to have been, *"You see, this is how you die."* Her Mercedes-Benz 600 survived and was much later acquired by the pop singer Jamiroquai. What would Coco have thought of that?

ELVIS
PRESLEY

1977 starb der King of Rock 'n' Roll, gerade mal 42 Jahre alt und längst unsterblich. Mit eigenwilligem, ja aufreizendem Hüftschwung, Schmalztolle, Koteletten und perfektem Gesang inszenierte das Talent aus Tennessee seine berühmten Songs, die bis heute unvergessen sind. Als einziger Künstler der Welt ist er in fünf Halls of Fame vertreten. „*I want to entertain people. That's my whole life. To my last breath.*" Presley wurde durch seine Live-Auftritte bald auch zum King of Las Vegas. Aber er blieb auf dem Boden: „*I am not the King. Jesus Christ is the King. I'm just an entertainer.*" Und er war Besitzer eines Mercedes-Benz 600: „*Ambition is a dream with a V8 engine.*"

The King of Rock 'n' Roll may have died in 1977 at the mere age of 42, but his legacy will live on forever. The talented Tennessean made the most of his spirited and downright provocative swivelling hips, pompadour, sideburns and perfect vocals to sing famous songs that have remained unforgettable. He is the world's only artist to have earned a place in five different halls of fame. *"I want to entertain people. That's my whole life. To my last breath."* Presley's live shows rapidly made him the king of Las Vegas. But he remained down-to-earth: *"I am not the King. Jesus Christ is the King. I'm just an entertainer."* He was the proud owner of a Mercedes-Benz 600: *"Ambition is a dream with a V8 engine."*

King of Rock 'n' Roll 1935-1977
Geliebt. Vermisst. Vergöttert.
King of Rock 'n' Roll, 1935-1977
Beloved. Mourned. Idolised.
"I'VE COME TOO FAR, AND I DON'T KNOW HOW TO GET BACK."
ELVIS PRESLEY

MERCEDES-BENZ
PAGODE
DESIGNIKONE SEIT 50 JAHREN

MERCEDES-BENZ
PAGODA
A DESIGN ICON FOR 50 YEARS

2.

DAS BERÜHMTESTE AUTODACH DER WELT

THE WORLD'S MOST FAMOUS CAR ROOF

ZURÜCK IN DIE ZUKUNFT – DER MERCEDES-BENZ 230 SL „PAGODE“ SCHAFFTE ALS ERSTER „MERCEDES“ DEN SPAGAT ZWISCHEN LIMOUSINE UND SPORTWAGEN.

RETRO-FUTURISTIC – THE MERCEDES-BENZ 230 SL “PAGODA” WAS THE FIRST MERCEDES TO DOUBLE AS A SALOON AND A SPORTS CAR.

828
Z-9424

Der Mercedes-Benz 230 SL (W 113) erschien 1963 mit ungewohnten neuen Proportionen und Linien – und dem unverwechselbaren „Pagodendach".
Debuting in 1963, the Mercedes-Benz 230 SL (W 113) boasted unconventional new proportions and contours – and an unmistakable "pagoda" roof.

1963 bekannte sich John F. Kennedy als Berliner, Jim Clark wurde Formel-1-Weltmeister und England wurde Zeuge des berühmtesten Postzugraubs der Geschichte. Deutschland erlebte zu den ersten Gehversuchen der Beatles und der Rolling Stones das Wirtschaftswunder. Das wiederkehrende Lebensgefühl ruft nach endlich lustvollen und anspruchsvolleren Automobilen. Einer der Stars der damaligen Zeit hat bis heute nichts von seinem Glanz verloren – der Mercedes-Benz 230 SL mit seinem nach innen gewölbten Hardtop. Die „Pagode" schreibt Geschichte. Bis heute.

Sie hießen Adenauer, Heckflosse oder Ponton. Aber auch Flügeltürer, Strich Acht, Flachkühler oder Baby-Benz. Zugegeben, die Spitznamen berühmter früherer Mercedes-Benz Modelle und auch die anderer Automobilmarken waren nicht von betont respektvoller Art. Dass auch den 1963 auf dem Genfer Automobil-Salon vorgestellten Mercedes-Benz 230 SL (intern W 113 genannt) dieses Schicksal ereilen würde, war von vorneherein klar und wäre auch nicht weiter erwähnenswert gewesen. Doch bei diesem Fahrzeug, das bald nur noch den Namen „Pagode" trug und dem selbst die französische Zeitung L'Equipe mit dem Slogan „Style Pagode" Respekt zollte, war alles anders: neuer, frischer, gewagter. Dazu ein bisschen französische Attitüde. Und eben dieses Dach! Was macht es bloß? Was aus dem Betrachter, was mit seinem Fahrer?

Bauwerklich betrachtet geht das auffällige Dach fernöstlich dominierter Pagodentempel schwungvoll und oft mehrgeschossig in die Höhe. Dem buddhistischen Glauben nach dient die Pagode der Aufbewahrung von Schätzen, dem Erhalt von Glück – oder auch ganz simpel als Aussichtsplattform. Von hinten betrachtet thront das leicht nach unten gewölbte Hardtop dieses Mercedes-Benz tatsächlich auf der Karosserie, wie ein eigenständiges Bauteil, ein Pavillon mit mit 360°-Panorama. Es ist eine Form, die beim ersten Anblick irritiert, das Auge beschäftigt. Der gläserne Aufbau spannt sich dazu über den Rest des Roadsters, dominiert ihn, polarisiert.

Diese knackige und für damalige Verhältnisse strenge und sachliche Formensprache spaltete die Kundschaft der Sechziger-Jahre. Das erforderte von Mercedes-Benz viel Mut, denn so sahen Autos damals normalerweise nicht aus. Heute auch nicht.

Das Pagodendach ist, und das unterstreicht die Faszination dieses SLs abermals, eine Erfindung von Béla Barényi, einem der größten Vordenker und Sicherheitsexperten des 20. Jahrhunderts. Gemeinsam mit dem genialen Karosseriedesigner Paul Bracq, der aus Barényis Idee eine lebendige konkrete Form kreierte, ließ er die bis dato völlig unbekannte Dachbauweise patentieren.

Kaum ein ernsthafter Automobilliebhaber kann sich noch heute der Faszination des im Volksmund auch schlicht als „der" SL bekannt gewordenen Sportwagens entziehen. Der erste seines Namens war er allerdings nicht: Er trug die Gene von 300 SL und 190 SL in

1963. John F. Kennedy gives his "Ich bin ein Berliner" speech, Jim Clark becomes Formula One World Champion, and the UK is rocked by the Great Train Robbery. As the German Wirtschaftswunder continues unabated, the Beatles and Rolling Stones are on the threshold of superstardom. There's a renewed sense of "joie de vivre", and people are finally starting to want cars that are more fun, more upmarket. One of the stars of this era has lost nothing of its luster today: the "new" Mercedes-Benz 230 SL, famous for its concave roof. The "Pagoda", as the car was nicknamed, made history – and continues to do so today.

"Adenauer", "Fintail", "Ponton", "Gullwing", „/8", „Baby Benz". It's fair to say the nicknames consumers gave to early Mercedes-Benz models – as well as models by other automakers – weren't necessarily respectful. It was clear that the Mercedes-Benz 230 SL (known internally as the W 113) would share this fate the moment it was unveiled at the Geneva International Motor Show in 1963. But with this car, soon known to everyone as the "Pagoda" – a nickname popularized by French newspaper L'Equipe – everything was different: it was newer, fresher, and more daring than the competition. It had a smidgen of French attitude. And this roof! What was its purpose? How would drivers, and onlookers, react to it?

The Buddhist pagoda temples of East Asia spear the sky dramatically with their architecturally striking roofs, often extending to multiple floors. In Buddhism, pagodas are used to store things of value, to preserve good fortune, and simply to act as viewing platforms. Seen from behind, the slightly concave hardtop of this Mercedes-Benz looks like a crown resting on top of the body – like a separate component, a miniature pavilion boasting panoramic views. It's a shape that bemuses and polarizes at first, one that demands a second look. The glass structure thrones over the whole of the roadster, dominating it.

Severe and matter-of-fact compared to other models of the era, the car's distinctive design divided customers. It was a courageous move by Mercedes-Benz – cars simply didn't look like that at the time. They still don't today.

The pagoda roof was invented by Béla Barényi, one of the automotive world's most pioneering 20th-century thinkers and a leading safety expert. He and body designer Paul Bracq, who translated Barényi's idea into vibrant reality, patented this completely novel roof design.

Today, this sports car has a mythical status that no serious car-lover can resist. Although popularly known as "the SL", it wasn't the first model to bear the name. The Pagoda inherited much of its build from its predecessors, the 300 SL and 190 SL, but looked much more suitable for everyday use than either. This must-have, must-drive model has now been working its magic for over fifty years – and still has just as much charm today.
By the time the last W 113 rolled out of production eight years later in March 1971, the

28c
S-PM 837

Ihre einzigartige Linienführung verknüpft die „Pagode“ mit dem zeitlosen Design der damaligen Architektur.
With its unique lines, the Pagoda is an automotive cousin of the timeless architecture of its period.

BELIEBT BEI DEN GROSSEN DIVEN: DORIS DAY, SOPHIA LOREN UND SENTA BERGER FUHREN „THAT WONDERFUL GERMAN CAR“ AUCH PRIVAT.

STARS LIKE DORIS DAY, SOPHIA LOREN AND SENTA BERGER DROVE THE “WONDERFUL GERMAN CAR” ONSCREEN AND OFF.

sich, trat jedoch alltagstauglicher als diese auf. Seit über fünfzig Jahren zieht der Sportwagen nun ungehemmt in seinen Bann, löst mit unverminderter Präsenz dieses Gefühl von must have und must drive aus.

Als der W 113 dann im März 1971, acht Jahre später, zum letzten Mal vom Band läuft, ist er längst unsterblich geworden. Bis dahin hat es die erfolgreiche Baureihe auf knapp 49.000 Exemplare gebracht – 34.115 wurden exportiert, davon allein 19.440 in die USA. Am 1. Oktober 1970 fragte Janis Joplin in einem ihrer berühmtesten Songs, drei Tage vor ihrem Tod, *„Oh Lord, won't you buy me a Mercedes-Benz“*. Sie wird bestimmt diesen SL gemeint haben…

Die bald Boulevard-erprobten, auf breite Spur gestellten und auffallend radbetonten Typen hießen 230 SL, später 250 SL, zuletzt 280 SL. Es gab sie als Coupé mit abnehmbarem Hardtop-Dach oder mit Roadster-Verdeck. Zur Verwandlung vom geschlossenen zum offenen SL mussten lediglich vier Hebel umgelegt werden, um das feste Pagodendach einfach abheben zu können. Doch die unverwechselbare Dachform hatte auch sehr pragmatische Motive: Sie sollte schlicht das Einsteigen in den recht tief bauenden und nicht sonderlich großen Sportwagen erleichtern: Kopf einziehen oder Nacken verrenken war kaum nötig. Hinzu kam eine überragende Rundumsicht dank der großen Fensterflächen. Trotz seiner optischen Leichtigkeit erhöhte die vorgespannte Dachform die Karosseriestabilität enorm – was vor fünfzig Jahren von entscheidender Bedeutung war, denn die damaligen Karosserien dürfen nicht mit den ausgetüftelten, bis ins kleinste Detail computerberech-

Pagoda had long since become immortal. Almost 49,000 vehicles had been manufactured – 34,115 of which were exported, a massive 19,440 of them to the U.S.. On October 1, 1970, in one of her most famous songs – recorded three days before her death – Janis Joplin asked, *“Oh Lord, won't you buy me a Mercedes-Benz?”*. She's sure to have meant this SL…

Three Pagoda models were created, all boasting a wide track and a wheel-focused design in addition to the characteristic roof shape. The first was the 230 SL, then came the 250 SL, and finally the 280 SL. They were available as coupés with a removable hardtop roof or a roadster-style soft-top. To take the pagoda roof off, all you had to do was release four catches holding it in place, then simply lift it off by hand. There were also pragmatic reasons for the roof's unmistakable shape: as the sports car had a low build and was not particularly large, the high roof aimed to make getting in and out easier. You barely needed to duck. The large windows also provided superball-round visibility. And though it looked light, the prestressed roof massively increased body stability, something that was of key importance fifty years ago, when the precisely computer-calculated bodies of today's vehicles were barely imaginable. The slightly concave roof also had other, more mundane practical benefits: you could set items down on it without them rolling off. Stars like Doris Day, Sophia Loren and Senta Berger drove the “wonderful German car” onscreen and off, often setting their lipstick or handbag down on the roof of the parked car.
Here's to fifty years of the Pagoda. The fact that this roof shape was retired, never gained

Das Design des SL, das neuer, frischer und gewagter war als alles bisher Dagewesene, wurde als „Style Pagode“ verewigt.
Newer, fresher and more daring than anything seen before, the SL's design was immortalised as "Pagoda style".

Glatte Flächen, viel Nutzraum und der Verzicht auf übermäßigen Zierrat zeichnen den luxuriös-puristischen „Pagoden-SL" aus. Insgesamt wurden knapp 49.000 Exemplare verkauft.
Smooth surfaces, lots of usable space and no extraneous embellishments: it can only be the luxurious, minimalist Pagoda SL. Just under 49,000 of these automobiles were sold in total.

neten Karosserien heutiger Automobile verglichen werden. Das leicht negativ gewölbte Dach war – aus damaliger Sicht – zudem von ganz anderem praktischen Vorteil: Gegenstände ließen sich dort ablegen, ohne Gefahr zu laufen, auf die Straße zu rollen. Zum Beispiel die Lippenstifte und Handtaschen von Doris Day, Sophia Loren oder auch Senta Berger. Im Film und auch privat fuhren die Diven zu gern „that wonderful german car".

Fünfzig Jahre Pagode also. Dass diese Dachform, von wenigen zaghaften Ansätzen auch bei anderen Fahrzeugherstellern abgesehen, dem W 113 vorbehalten und den Nachfolgern versagt blieb, hatte vor allem aero-dynamische Gründe: Der Luftwiderstandsbeiwert des Hardtops betrug abenteuerliche 0,52 – das Doppelte des heute Üblichen.

Kraftstoffverbrauch, Emissionsvorschriften und Windgeräusche lassen das Pagodendach in seiner Urform in unserer Zeit unrealistisch erscheinen. Was letztendlich das wohl schönste Dach der Automobilgeschichte erst recht unsterblich macht. •

broad popularity and was never adopted by other models, apart from a few cautious attempts by other carmakers, was primarily due to aerodynamic reasons. The hardtop had an eyebrow-raising drag coefficient of 0.52 – twice as high as is commonplace today. Fifty years on, in a world where low fuel consumption, emissions regulations and low wind noise are all prime concerns, the original pagoda roof – probably the most beautiful roof in automotive history – seems unrealistic and impractical. But it's precisely this that makes it truly immortal. •

Eine Sonderausführung des 250 SL ohne Roadsterverdeck – dafür mit Rücksitzbank – aus dem Jahre 1967, das California-Modell.
The California Coupé, a special edition of the 250 SL from 1967, lacked a roadster soft-top but included a rear bench seat.

Hochwertige Materialien und perfekte Verarbeitung kennzeichneten den Innenraum und gaben Möglichkeiten zur individuellen Gestaltung. Das lederähnliche „MB-Tex“ war Serie. Optional gab es etwa Lederpolsterung oder das berühmte Hardtop mit der Delle.

Premium materials and painstaking craftsmanship define the car's interior. There were also customised design options. The leather-effect MB-Tex material was the series standard. Buyers also had the choice of leather upholstry or the famous concave hardtop.

Lieferbar war der SL in mehreren Ausführungen: als Roadster mit Klappverdeck, als Coupé mit abnehmbaren Dach oder mit Roadsterverdeck.
The SL was available in a variety of version: as a Roadster with a convertible top, and as a Coupé with a removable roof or a roadster soft-top.

[redacted]

Sonderausstattungen, die es in sich hatten. Halogenlichter, Servolenkung sowie ein 4-Gang-Automatikgetriebe machten das Fahren im „Pagoden-SL“ komfortabler und sicherer als in vielen anderen Sportwagen.
Special editions that lived up to the name. Halogen lights, power steering and 4-gear automatic transmission made driving the Pagoda SL safer and more pleasant than many other sports cars.

MASKULIN OHNE ZU SEHR MIT DEN MUSKELN ZU SPIELEN. WEICH OHNE DIE MASSGEBLICHEN KONTUREN UND KLAREN LINIEN ZU VERLIEREN. MIT DEM „PAGODEN-SL“ ERFAND MERCEDES-BENZ SPORTLICHES OFFEN FAHREN NEU.

EFFORTLESSLY MASCULINE WITHOUT BEING MACHO. SOFT WITHOUT LOSING ITS CHARACTERISTIC STRAIGHT LINES AND CONTOURS. IN THE FORM OF THE PAGODA SL, MERCEDES-BENZ TOTALLY RETHOUGHT SPORTY, OPEN-TOP DRIVING.

220 S
D
BB CZ 79

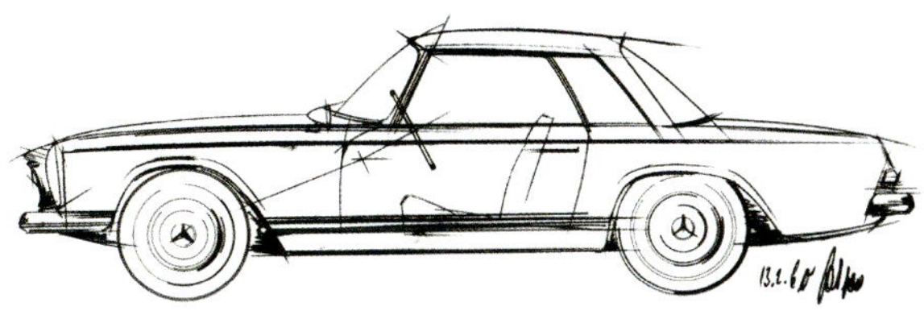

PAUL BRACQ

MAÎTRE CAROSSIER

DIESES DACH! ES MACHTE AUS DEM 230 SL EINE DER BERÜHMTESTEN AUTOMOBILEN SKULPTUREN – DIE PAGODE. WAR ES DAS ALLEINIGE WERK DES AUTOMOBIL-DESIGNERS PAUL BRACQ? MIT DEM EBENSO GENIALEN WIE BODENSTÄNDIGEN FRANZÖSISCHEN MEISTER DREHEN WIR DIE UHR UM FÜNFZIG JAHRE ZURÜCK. UND STAUNEN.

THAT ROOF! THE “PAGODA” TURNED THE 230 SL INTO ONE OF THE MOST FAMOUS AUTOMOTIVE SILHOUETTES. WAS DESIGNER PAUL BRACQ THE SOLE MASTERMIND BEHIND THIS STRIKING STYLE? WE SAT DOWN AND TURNED THE CLOCK BACK FIFTY YEARS FOR AN ASTONISHING CONVERSATION WITH THE FRENCH MASTER, WHO IS AS BRILLIANT AS HE IS DOWN-TO-EARTH.

Vorherige Seite: Béla Barényis Mercedes-Benz Typ 220 S „Versuchsträger"; Paul Bracq (kniend) beim Belastungstest auf der Versuchsbahn in Untertürkheim, 1965.
Previous page: Béla Barényi's 220 S Mercedes-Benz "test case"; Paul Bracq (kneeling) during a stress test on the test course in Untertürkheim, 1965.

Weltpremiere auf dem Genfer Automobil-Salon 1963: Paul Bracq mit Béla Barényi neben dem neuen 230 SL.
World premiere at the Geneva Automobile Salon in 1963: Paul Bracq with Béla Barényi next to the new 230 SL.

Paul Bracq in seinem Atelier am Zeichenbrett.
Paul Bracq at the drafting table in his studio.

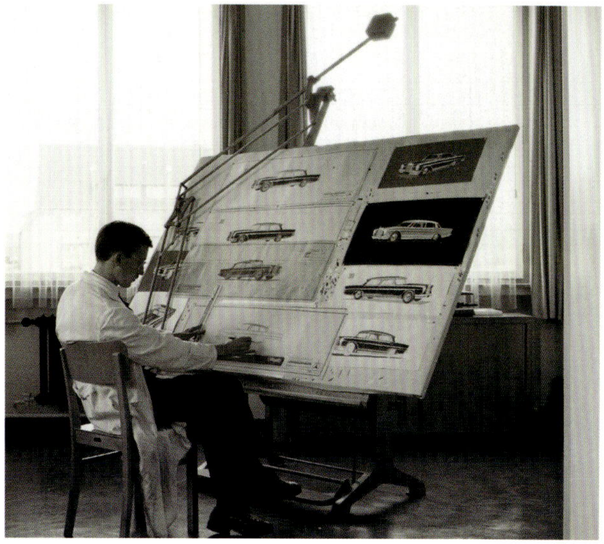

Wie alle Genies winkt er ab, wenn er auf seinen größten Wurf angesprochen wird. *„Non Monsieur - nicht ich habe das Pagodendach erfunden."* Paul Bracq treibt die Bescheidenheit auf die Spitze, will nicht einmal als „Designer des Pagodendachs" bezeichnet werden. *„Und ich bin auch kein Designeur"*, beharrt er mit starkem französischem Akzent, den die Deutschen so lieben.
Warum diese Selbstverleugnung? Bracq ist nämlich durchaus ein Designer, ein Automobildesigner, genauer gesagt. Und einer der berühmtesten dazu. Zwar ist sein Name selbst unter Insidern nicht so präsent wie ein Bertone, Pininfarina oder Giugiaro. Gäbe es eine Hall of Fame der Karosserie-Couturiers, so wäre sein Name für immer in Marmor gemeißelt.

Drei lange Jahre, die seinen erwachenden Esprit unterbrachen, aber gewiss nicht dämpften.

Noch heute, mit achtzig Jahren, lässt Bracq seinem Temperament freien Lauf, wenn er betont: *„Ich bin Carossier, ein Karosserie-Gestalter! Ich kämpfe für die Kunst der Karosserie-Gestaltung!"* Eine ebenso klare Antwort gibt der Maître Carossier auf die Frage, wer denn nun das legendäre Pagoden-Dach des Mercedes-Benz 230 SL erfunden hat - eine Frage, die er schon viel zu oft gehört hat, die ihm dieses eigenartige Unbehagen bereitet. Doch vielleicht kokettiert er nur ein wenig mit seiner eigenen unantastbaren Vita, wenn er sagt: *„Es war eine Teamarbeit von Karl Wilfert als Chef des Bereichs Fahrzeugkonstruktion und Vorentwicklung, Béla Barényi als Vordenker und mir. Doch die ursprüngliche Idee stammt von Barényi. Als ich im Jahr 1957 nach meinem Militärdienst zu Mercedes-Benz kam, arbeitete ich bei Béla Barényi in der Fahrzeug-Vorentwicklung. Meine erste Aufgabe bestand darin, das Design einer von Barényi mit identischen vorderen und hinteren Kotflügeln symmetrisch gestalteten Automobil-Studie zu verfeinern. Es gab bis dahin nur Handskizzen von Barényi."* Der 1907 geborene Österreicher war als Konstrukteur bei Mercedes-Benz seiner Zeit weit voraus, gilt bis heute als

„ICH KÄMPFE FÜR DIE KUNST DER KAROSSERIE-GESTALTUNG!"

Am 13. Dezember 1933 erblickte Paul Bracq in Bordeaux das Licht der Welt. Er besuchte von 1950 bis 1953 die renommierte Pariser Design- und Kunstschule École Boulle, heuerte nach seinem Abschluss beim Designer Philippe Charbonneaux an, dessen Werke später zu festen Institutionen im Centre Pompidou in Paris oder im Museum of Modern Art in New York wurden. Diese erste wichtige Station währte allerdings nur kurz, denn 1954 musste er seinen Wehrdienst bei der französischen Armee bis 1957 leisten.

Like all geniuses, he waves his hand dismissively when asked about his greatest achievement: *"Non, monsieur – I did not invent the Pagoda roof"*, Paul Bracq takes his modesty to extremes – he does not even want to be known for the Pagoda design. *"I am also not a designer,"* he says with his charmingly thick French accent.
Why such self-effacement? After all, he is most certainly a designer – an automotive designer, to be more specific, and one of the most famous ever. It may well be true that Bracq is not as well-known to insiders as other luminaries such as Bertone, Pininfarina or Giugiaro. But if there were a Hall of Fame for automotive body couturiers, his name would certainly be chiselled in marble.

Paul Bracq was born on 13 December 1933 in Bordeaux.He attended the renowned École Boulle School of Design and Art in Paris from 1950 to 1953, and after graduation he started working for designer Philippe Charbonneaux, whose works later became veritable institutions at places such as the Centre Pompidou in Paris and the Museum of Modern Art in New York. This formative phase of his early career did not last long, however, since he was enlisted and served in the French army from 1954 to 1957.

These three long years interrupted his growing sense of esprit but certainly did not dampen it.

Now 80, Bracq's temperament is still clearly evident when he emphatically states, "I am a carossier, an automotive body stylist! I am dedicated to the art of body styling!" The maître carossier answers just as clearly when asked who designed the legendary Pagoda roof of the Mercedes-Benz 230 SL. He has heard the question too many times, and it gives him an unusual sense of discomfort. But perhaps he is also being coquettish about his exceptional background when he responds by saying, *"The roof was the result of some excellent teamwork between Karl Wilfert as Head of the Technical Design and Predevelopment department, Béla Barényi with the idea that inspired it, and myself. But the original idea came from Barényi. When I came to Mercedes-Benz in 1957, after completing my military service, I worked under Béla Barényi in vehicle predevelopment. My very first assignment was to refine the lines of a symmetrical automobile study designed by Barényi with identical front and rear wings. At that point we only had his hand-drawn sketches to work with."* Born in Austria in 1907, Barényi did work as an engineer at Mercedes-Benz that was years ahead of its time,

"I AM A CAROSSIER, AN AUTOMOTIVE BODY STYLIST! I AM DEDICATED TO THE ART OF BODY STYLING!"

Designstudien wie der K-55 gaben wichtige Rückschlüsse auf das Zusammenspiel von Farbe und Form. Auch bei dieser Studie erkennt man schon deutlich, dass das Dach ein wichtiges Element der Konstruktion war.
Design studies such as K-55 revealed important information about how contours and colours interacted. The study also clearly showed how vital the roof was as a design element.

DAS KONKAVE DACH, ES LIESS BRACQ EBENSO WIE BARÉNYI UND WILFERT NICHT MEHR LOS.

Begründer der passiven Sicherheit von Automobilen. Barényi hatte mit der Studie, die er Bracq überließ, Spezielles im Sinn. Er plante einen Allzweckwagen, einen für alles.

„Das bei dieser Studie konkav, sprich: nach innen gewölbte Dach, wies durch diese ungewöhnliche Gestalt eine große Tragfähigkeit auf und sollte unseren Prototyp zum Multifunktions-Automobil machen. Die Dachfläche wollte Barényi als Nutzfläche gestalten. Für ein Campingzelt, ja, als Aussichtsplattform, als Stehfläche für Fernsehreporter, als Hochsitz für die Tigerjagd. Was auch immer.“
Die praktische Erprobung erfolgte in Sindelfingen mit einigen Ponton- und Heckflossen-Limousinen, deren Dächer konkave Aufsätze erhielten. Aus dem Allzweckwagen wurde nie etwas, die Dachkonstruktion hingegen fand ihren Weg in ein damals aktuelles Projekt – den W 113, der später nur noch „die Pagode“ genannt wurde. Bracq hatte, ohne es in seiner ganzen Bedeutung zu erfassen, einen Meilenstein geschaffen, einen Obelisk des Karosseriedesigns. Ob er später selbst überrascht gewesen sei? Bracq schweigt vielsagend.

Das nach innen gewölbte Dach war aber nicht nur besonders tragfähig. Es machte zudem den ganzen Aufbau steifer und formstabiler. Das faszinierte den damals vierundzwanzig Jahre jungen Chef der Abteilung Advanced Design. Sein erster großer Auftrag war gleich sein bedeutendster: ein neues Coupé-Dach mit vergrößerter Heckscheibe für den 190 SL zu gestalten. Inzwischen zum Leiter der Design-Vorentwicklung avanciert, schuf Bracq also ein elegantes Hardtop, das im Herbst 1959 das plötzlich altbacken wirkende frühere Dach ablöste. Und als der junge französische Designer noch im selben Jahr den für 1963 geplanten Nachfolger des 190 SL vorskizzierte, verwendete er zunächst diesen Coupé-Gedanken weiter. Bracq war in seinem Element, er glättete die Flanken des Sportwagens, modulierte die Heckpartie sehr flächig. Ließ sich bewusst nicht von den in Mode geratenen Heckflossen-Limousinen amerikanischer Prägung beeinflussen. Das war nicht sein Stil. Und sein Glück. Denn dass man später einmal ebenso deutlich wie bewundernd sagen würde, ein Franzose habe die formschönsten deutschen Automobile entworfen, hätte er sicher nicht geglaubt.

Das konkave Dach, es ließ Bracq ebenso wie Barényi und Wilfert nicht mehr los. Es war 1960, als das Trio es beim 230 SL in die Realität umzusetzen begann. Schließlich war Paul Bracq war derjenige, der die von Béla Barényi im Grundsatz erdachte Dachform zeichnerisch ebenso elegant wie harmonisch und stilsicher in Form goss.
Aus damaliger Sicht bot das konkave Dach

THE IDEA OF A CONCAVE ROOF CONTINUED TO FASCINATE BRACQ, BARÉNYI AND WILFERT.

and to this day he is seen as the founding father of passive automotive safety. Barényi had something special in mind with the study he gave Bracq: he was planning an all-purpose car.

"The concave roof of this study vehicle had a large carrying capacity thanks to its dished structure, and was intended to make the prototype a multifunctional automobile. Barényi wanted the roof surface to function as a load-carrying structure – for a camping tent, as a viewing platform, as a standing area for television reporters or as a vantage point for tiger-hunting, for example." Practical tests were conducted in Sindelfingen, using a number of Ponton and Fintail saloons whose roofs had been fitted with concave superstructures. The all-purpose vehicle never worked out, but the roof design made its way into another project at the time – the W 113, which was later known simply as the Pagoda. Without having fully realised what he had achieved, Bracq created a benchmark and set a milestone in body design. Does this fact surprise him? Bracq's silence speaks volumes.

The roof with a slight inward curve not only had an enhanced carrying capacity: it also made the entire structure of the car more rigid and stable. This intrigued the then 24-year-old head of the Advanced Design department. His first major assignment turned out to be his most significant: creating a new coupé roof with a larger rear window for the 190 SL. Soon afterward Bracq was promoted to head of Styling Design Predevelopment, and there he created an elegant hardtop as a successor to the previous roof which had grown stodgy-looking by the fall of 1959. And later that year, he held onto the idea of the coupé roof as he started design work for the 190 SL which was scheduled for release in 1963. Bracq was in his element: he smoothed the sport car's flanks and gave the rear section plain surfaces. He deliberately avoided the influence of the American-type fintail saloons which were in fashion– that was not his style, and thank heavens it wasn't. He would never have believed that people would later say with equal amounts of conviction and admiration that a Frenchman designed Germany's most stylish automotive silhouette.

The idea of a concave roof continued to fascinate Bracq just as it did Barényi and Wilfert. It was not until 1960 that the trio started to turn their plans into reality with the 230 SL. And Paul Bracq was the one who translated the ideas in Béla Barényi's sketches into a style that was as elegant as it was harmonious and confident.
At the time, the concave roof was thought to offer advantages that not only affected the car's style but its construction as well: higher

Stilikone - nicht nur außen. Formen und Farben des Interieurs sind Ausdruck einer neu gedachten Persönlichkeit. Auffallend die filigranen Details, das Spiel mit Farbtönen, die Verkleidung der Armaturentafel und das Fehlen von Ecken und Kanten im gesamten Innenraum.
The exterior was not the only thing with an iconic style. The lines and colours of the interior expressed a different vision of a personality – the exquisite details, the interplay between different shades, the panelling on the dashboard, the lack of corners and angles throughout the interior.

nicht nur stilistische, sondern vor allem konstruktiv überragende Vorteile: höherer Einstieg, größere Glasflächen und dadurch bessere Rundumsicht sowie verbesserte Karosseriestabilität. Heute jedoch, fünfzig Jahre später, werden die Vorteile des Pagodendachs in der Fachliteratur gelegentlich angezweifelt. Das ist der Punkt, an dem Paul Bracq zum Kämpfer wird, zum Advocat de la Pagode: *„Wenn wir ein klassisch nach oben gewölbtes Dach entworfen hätten, wäre es bei gleicher Stabilität und gleich hohem Einstieg zehn Zentimeter höher geraten und hätte das Auto zu bullig wirken lassen. Die Alternative zum konkaven Dach wären zwangsläufig eben der um zehn Zentimeter verringerte Einstieg und die schlechtere Rundumsicht gewesen. Das Pagodendach hat also in erster Linie funktionelle Gründe! Die Ästhetik und der moderne Look spielten die zweite Geige."* Bracq atmet durch, tief. Immer wieder muss er diese Umstände klarstellen. Und er setzt nach: *„Außerdem wäre dann der W 113 vielleicht nie zu jenem begehrten Klassiker geworden, der er heute ist."*

Dabei beruft sich der Maître auf Versuche mit Schattenwürfen eines von innen ausgeleuchteten Pagoden-Cockpits an den Wänden eines verdunkelten, mattschwarz gehaltenen Raums. Der optische Effekt verblüffte Besucher immer wieder: Bei der Pagode haben nur die A-Säule der Frontscheibe und die C-Säule zwischen Seiten- und Heckscheibe tragende Funktion. Diese beiden Pfosten sowie das Dach und die Flanken bleiben bei der Wandprojektion dunkel. Die Glasflächen aber erscheinen hell auf der Wand, so lassen sich Fensterflächenanteil und Rundumsicht gut ermitteln. Großes Kino.

Es lässt sich nicht leugnen, die Pagodenform hatte auch unbestreitbare Nachteile. Betrachten wir dieses außergewöhnliche Auto mit heutigen Augen, so heißt der schärfste Wider-

entry clearance, larger window surfaces for better all-round visibility, and a more stable body. Fifty years later, however, specialists sometimes question the benefits of the Pagoda roof. And this is where Paul Bracq becomes a warrior, an "avocat de la Pagode": *"If we had designed a convex roof along conventional lines, it would have needed to be ten centimetres higher to obtain the same stability and entry clearance, and that would have made the car appear top-heavy. The alternative to a concave roof would necessarily have been a ten-centimetre lower entry clearance and poorer all-round visibility. The Pagoda roof concept was therefore chosen primarily for functional reasons. Visual appeal and a modern look, though important, were of secondary concern."* Bracq takes a deep breath. He always has to clarify this issue. He adds, "Besides, the W 113 might not have gone on to become the sought-after classic it is now."

Paul Bracq discusses the practical tests that were conducted: they put the Pagoda in a darkened room with matte black walls and used a powerful lamp to illuminate the interior of the cockpit so that they could see where shadows were projected onto the walls. The optical effect is stunning every time. In the Pagoda, the A-pillar at the windscreen and the C-pillar between the side windows and rear window are the only load-bearing areas that support the roof, and when you light the interior from within, the dark shadows on the walls only come from these two pillars, the roof itself and the vehicle flanks. By contrast, the areas corresponding to the windows appear in bright light from all sides when projected onto the wall. It is striking to see the relative surfaces of the windows and excellent all-round visibility.

In the meantime it is now clear to us that the pagoda shape has undeniable disadvantages.

"ICH BIN KEIN DESIGNER, SONDERN CAROSSIER, KAROSSERIE-GESTALTER!"

"I AM NOT A DESIGNER – I AM A CAROSSIER, AN AUTOMOTIVE BODY STYLIST!"

Auf der Suche nach neuen Formen: Zeichnung zum Projekt K-55 von Paul Bracq, 1964.
The quest for a new shape: drawing by Paul Bracq for the K-55 project, 1964.

[redacted]

Paul Bracq an einer Studie zum späteren Experimentalfahrzeug C 111, das 1969 auf der IAA Weltpremiere feiert.
Paul Bracq and a model of the yet-to-be-built C 111 experimental vehicle, later premiered at the Frankfurt Motor Show in 1969.

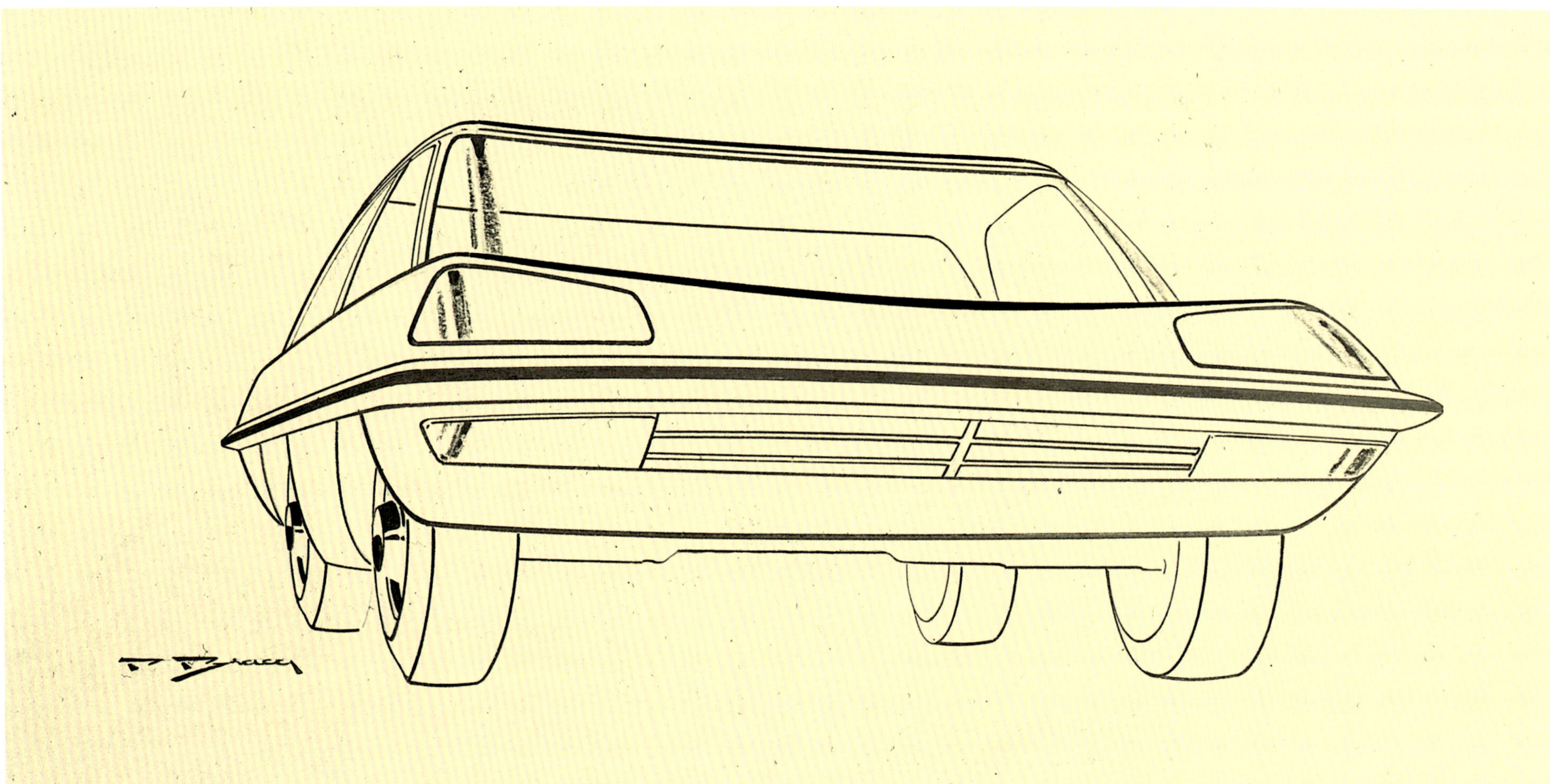

sacher Aerodynamik. Die Fakten liegen schonungslos auf dem Tisch: Gegenüber dem 190 SL mit einem Luftwiderstandsbeiwert von 0,461 verschlechterte sich der cw-Wert des 230 SL mit aufgesetztem Pagodendach auf 0,515 - mit geschlossenem Stoffdach betrug er immer noch 0,481. Doch gedanklich sind wir im Jahr 1963, eine nicht perfekte Luftumströmung im Detail interessierte damals weniger, auch nicht der daraus resultierende erhöhte Kraftstoffverbrauch. Das waren Feinheiten, die man noch ignorieren oder zu Gunsten anderer Ziele opfern durfte. Damals.

Ungeachtet dessen machte die faszinierende Dachform Karriere, besser gesagt: Furore. Sie schlug ein wie eine Sexbombe der Swinging Sixties, polarisierte, faszinierte, schreckte auf und ab. Sie war es, die dem Zweisitzer bereits während seiner Bauzeit im Volksmund zum Kosenamen Pagode verhalf. Eingebrannt bis heute im Gedächtnis. Solche Aufmerksamkeit, Liebe, oder kurz: Ehre wurde nur wenigen anderen Fahrzeugen in der Geschichte zuteil.
Die Hall of Fame, sie hat noch Platz. •

If we look at this unusual car through a modern perspective, its greatest problem is aerodynamics. And the facts stare you in the face: compared to the 190 SL with a drag coefficient of 0.461, the Cd-value of the 230 SL with a mounted Pagoda roof increased to 0.515, although with a closed fabric soft top the figure was only 0.481. But let us see this from the perspective of the year 1963, when the idea of suboptimal aerodynamics was not much of a concern, nor was the resulting increase in fuel consumption. Those were details that people ignored or sacrificed in favour of other features – back then, at least.

Nevertheless, the contoured roof not only became a household name, it created a furore. It blew up like a sex bomb in the „swinging sixties“ – polarising, fascinating, alluring and startling. It was what earned the two-seater the affectionate popular nickname of the Pagoda even while the car was still in production. It has remained seared into our memory to this very day. Very few other cars in history have generated the same kind of attention, love or even respect.
There's still room in the Hall of Fame. •

39

2.3

SIEGEN
AM LIMIT
VICTORY
AT THE EDGE

1963 GEWINNEN EUGEN BÖHRINGER UND SEIN CO-PILOT KLAUS KAISER IM 230 SL DIE RALLYE SPA – SOFIA – LIÈGE.

IN 1963 EUGEN BÖHRINGER AND HIS CO-PILOT KLAUS KAISER WON THE RALLY SPA – SOFIA – LIÈGE IN THEIR 230 SL.

230 SL im strapaziösen Rallyeeinsatz:
Mit Vollgas durch die Kurve.
The 230 SL during the exhausting rigours of the rally: taking a curve at top speed.

[redacted]

90 Stunden sind Fahrer und Beifahrer ununterbrochen „auf Achse“ – die Fahrt vom belgischen Spa bis nach Sofia und zurück nach Lüttich ist ein Kampf gegen miserable Straßen, schlechte Sicht, Staub, extreme Hitze, Geröll, Matsch und Müdigkeit. Dass von 107 gestarteten Autos am 31. August 1963 nur 21 in Lüttich ankommen, wundert nicht. Ganz vorne stehen am Ende Eugen Böhringer und sein Co-Pilot Klaus Kaiser. Für den am 19. Juni 2013 verstorbenen schwäbischen Rennfahrer und Mercedes-Benz Werkfahrer Eugen Böhringer war dies der zweite Sieg beim „Höllentrip“ – 1962 gewann er schon einmal, damals auf einer 220 SE „Heckflosse“.

Dabei war die Chefetage um den Vorstandsvorsitzenden Walter Hitzinger zunächst unsicher, ob es klug sei, den neuen 230 SL (W 113) tatsächlich für eine harte Rallye voller Unwägbarkeiten einzusetzen: Wenn das schief ginge, sei das schlecht für das Image eines Sportwagens, der sowieso ein schweres Erbe antreten würde. Schließlich sollte die „Pagode“ zugleich den 190 SL und den 300 SL Roadster ersetzen. Professor Dr. Fritz Nallinger, der damalige Chef der Mercedes-Benz Entwicklungsabteilung, ließ nicht locker: Er beantragte drei Testfahrzeuge, die er auf der sogenannten „Schlechtwetterstrecke“ des hauseigenen Versuchsgeländes malträtierte. Fahrer und Beifahrer machten sich parallel mit Schlafsack, Spirituskocher, Kochtopf und Serienfahrzeug auf und fuhren den Originalkurs ab. In das Rallye-Roadbook, auch „Gebetsbuch“ genannt, notierte Kaiser, was er Böhringer während der Rallye „vorzubeten“ hatte. Währenddessen wurde das Rallye-Auto nach heutigen Maßstäben moderat umgerüstet: Zum Einsatz kamen ein verstärktes Fahrwerk, Servolenkung und Unterfahrschutz. Der Motor des 230 SL bekam einen überarbeiteten Zylinderkopf, größere Kolben und wurde zusätzlich von einem Ölkühler auf Arbeitstemperatur gehalten. Das Auto fuhr – je nach Straßenverhältnissen – auf 13- bzw. 15-Zoll-Rädern.

Und wie der 230 SL fährt: Er wird von der Meute über eine Strecke von 5.500 Kilometern vor sich her gehetzt. Meist sitzt Böhringer am Steuer, sein Beifahrer übernimmt das Lenkrad auf den „besseren“ Straßen. Dann gönnt sich der damals 41-jährige Böhringer einen viertelstündigen „Powernap“. Gefahren warten überall auf „Pagode“ und Fahrer, zum Beispiel nachts auf dem über 2.600 Meter hohen Paso di Gavia in der Lombardei. Der Feldweg ist mit spitzen Steinen besät, nur zwei Meter breit, links und rechts geht es steil nach unten. Doch Böhringer, Kaiser und ihr 230 SL setzen sich durch. 1964 – ein weiterer Triumph für Böhringer: Im vorerst letzten Rallyejahr von Mercedes-Benz wird der Schwabe Dritter bei der strapaziösen Marathon-Rallye. Weil er vier Mal hintereinander im Spitzenfeld ins Ziel kommt, erhält der Rallye-Europameister von 1962 am Ende den verdienten „Goldpokal“. Diese Trophäe wird der glanzvolle Höhepunkt einer außergewöhnlichen Motorsport-Karriere des Mercedes-Benz Werksfahrers Eugen Böhringer. •

The drivers and co-drivers were en route for 90 hours non-stop: the race from the Belgian town of Spa to Sofia and back to Liège was a battle against miserable roads, poor visibility, dust, extreme heat, gravel, mud and fatigue. It was no surprise that 107 cars left Spa on 31 August 1963 but only 21 made it back to Liège. The men who ultimately crossed the finish line first were Eugen Böhringer and his co-driver Klaus Kaiser. Eugen Böhringer was a southern German racecar driver who worked as a Mercedes-Benz factory driver and died on 19 June 2013. The 1963 victory was his second – he‘d won the “trip through hell” once before in 1962 whilst driving a 220 SE Fintail.

For his second race, the board of management under the direction of CEO Walter Hitzinger was not sure whether it was wise to put the new 230 SL (W 113) to the test during a gruelling rally full of uncertainties. If it didn‘t go well, it would have been very bad for the image of a sports car that would be taking on a heavy burden anyway. After all, the „Pagoda“ was intended to replace both the 190 SL and 300 SL Roadster. Professor Dr. Fritz Nallinger, then the head of Mercedes-Benz’s development department, persisted: he took three test vehicles and put them through their paces on the area known as the “bad-weather stretch” of the company’s test track. In the meantime, the driver and co-driver headed out in tandem with a sleeping bag, spirit stove, a pot and a series-production vehicle to drive the original route. Kaiser followed the pacenotes, also known as the “prayer book”, to help Böhringer „say his prayers“ by closely following every step of the route. Back at the plant, the rally car was undergoing modifications that seem modest by today‘s standards: a reinforced chassis, power steering and underrun protection. The 230 SL’s motor was given a modified cylinder head, larger pistons and was cooled by an oil-water heat exchanger. Depending on road conditions, the car drove on 13- or 15-inch tyres.

And the 230 SL could drive: it dashed along with the rest of the pack along a stretch of 5,500 kilometres. Usually Böhringer sat at the wheel, and his passenger took over on some of the “better” roads so that Böhringer, then 41, could take a fifteen-minute “power nap”. For the Pagoda and its drivers, dangers were lurking everywhere, such as driving at night the along the Paso di Gavia in Lombardy at an altitude of over 2,600 metres. The dirt road there is lined with jagged stones and is a mere two metres wide, plunging straight down steeply on either side. But Böhringer, Kaiser and their 230 SL pressed on. 1964 brought another triumph for Böhringer: in what was to be Mercedes-Benz’s last rally year, the Swabian placed third in the exhausting marathon rally. Since he made it to the last staging of the rally four times in a row, the 1962 European Champion was awarded a well-earned gold trophy. This cup was the crowning achievement of the exceptional motorsport career of Mercedes-Benz factory driver Eugen Böhringer. •

S-RV 441

Der Mercedes-Benz 230 SL mit der Startnummer 39 vollbringt 1963 eine Glanzleistung: mit ihm gewinnt das Team um Werksfahrer Eugen Böhringer das anspruchsvolle Langstrecken-Rennen Spa-Sofia-Liège.

The Mercedes-Benz 230 SL that sported number 39 achieved glory in 1963, successfully leading the team under driver Eugen Böhringer to victory on the demanding Spa-Sofia-Liège open-road race.

39
SPA SOFIA LIEGE

DAS PAGODEN-RENNTEAM EUGEN BÖHRINGER UND KLAUS KAISER FREUEN SICH ÜBER IHREN GESAMTSIEG.

IN 1963 EUGEN BÖHRINGER AND HIS PAGODE WIN THE RALLYE SPA – SOFIA – LÜTTICH.

UNTERWEGS
GESCHICHTEN NAH AM ASPHALT
ON THE ROAD
TALES FROM THE ASPHALT

3.

300D
DIESEL
California
6MSD241

NEW GENERATION

LIFESTYLE MERCEDES YOUNGTIMER – EIN MODERNES ROADMOVIE.

THE MERCEDES-BENZ VINTAGE CLASSIC LIFESTYLE – A MODERN ROAD MOVIE.

J.G. Francis, Eigentümer von Mercedes Motoring, und Geschäftspartner Sean Johnstun sind immer auf der Suche nach neuen Fundstücken, die sie bis ins kleinste Detail liebevoll restaurieren.
J.G. Francis, who owns Mercedes Motoring, and his business partner Sean Johnstun are always on the lookout for new treasures they can lovingly and thoroughly restore top to bottom.

Es ist nicht leicht, Jimmy George Francis, kurz J.G., zu treffen. Weil er eigentlich immer unterwegs ist, meistens mit seinem Kumpel Sean Johnstun. Und weil den beiden Kaliforniern dann ständig irgendwelche unglaublichen Dinge passieren: sei es ohne Orientierung nachts in der mexikanischen Wüste umher zu irren, ohne Scheibenwischer in einen Schneesturm zu geraten oder sich bei einem High-Speed-Fahrwerktest, abseits der empfohlenen Routen, ein riesiges Loch in die Ölwanne zu reißen. Es ist also nie genau vorherzusagen, wann die beiden von einer ihrer Wahnsinnstouren zurück sind. Sicher ist nur, dass sie garantiert einen ihrer extrem coolen Vintage-Mercedes dabei haben.
Denn diese Liebhaberstücke suchen, finden und restaurieren J.G. und Sean seit 2003 aus Leidenschaft. Insbesondere Dieselmodelle der Baujahre 1968 bis 1985, die besten Autos aus Serienproduktion der Welt, ist J.G. überzeugt. Mehr als 300 dieser Classic Cars haben J.G. und Sean in den letzten elf Jahren mit J.G.'s Firma Mercedes Motoring restauriert. Work-Sessions in der Garage als Abschluss und Auftakt für die nächsten Abenteuer, über die die Jungs regelmäßig in ihrem Blog berichten.

„MERCEDES-BENZ DIESELMODELLE DER BAUJAHRE 1968 BIS 1985 SIND DIE BESTEN AUTOS DER WELT AUS SERIENPRODUKTION."

Zum Beispiel über die Reise mit einem eierschalenfarbenen 240 D Mercedes-Benz aus dem Jahr 1976, den sie in Alaska entdeckt, ungesehen gekauft und abgeholt hatten.
Frei nach J.G.'s Lieblingszitat von Hemingway: „Die kurze Antwort lautet: Tu es einfach!" Ein Sammler-Juwel mit nur 74.000 Originalkilometern auf der Uhr. Der perfekte Anlass für einen Roadtrip der Extraklasse von Alaska heim nach L.A.

Leseprobe aus J.G.'s Roadbook: „Alaska, 14. August – Tag 6. Wenn ich jemals darüber grübeln würde, dass ich nie genug Zeit oder

It's not easy to make a date with Jimmy George Francis, who calls himself J.G. – he's basically always on the road, usually with his friend Sean Johnstun. And the most amazing things tend to happen to the two Californians: they end up lost in the Mexican desert at night, find themselves in the middle of a blizzard without windshield wipers, or do a high-speed suspension test off-road and then rip a huge hole in their oil pan. So it's impossible to say when the two will return from one of their unbelievable trips. The only thing that's certain is that while they are on the road, they are guaranteed to be in one of their extremely cool vintage Mercedes.

"DIESELS BUILT BETWEEN 1968 AND 1985 ARE THE WORLD'S BEST MASS-PRODUCED CARS."

Since 2003, J.G. and Sean have put their heart and soul into tracking down and restoring these collector's items, especially diesels made between 1968 and 1985, which, according to J.G., are the world's finest mass-production cars. Mercedes Motoring, the company founded by J.G., has restored over 300 of these classic automobiles over the last eleven years. The work they do in the garage completes one project and simultaneously gets them ready for their next adventure, and the two regularly report about their travels in their blog. One example was their road trip in an eggshell-coloured 1976 240 D Mercedes-Benz they bought in Alaska sight unseen and drove back to L.A. – true to J.G's favourite Hemingway quote: "The shortest answer is doing the thing!" It was a magnificent collector's item with only 74,000 original kilometres under its belt. The perfect reason for a once-in-a-lifetime road trip from Alaska back to L.A.

A page from J.G.'s journal reads: "Alaska, August 14th – Day 6. If I worried about not having enough time or enough money to do all the things I ever wanted to do, I probably wouldn't have done anything with my life. Besides, there is never enough time and never enough money, so those are the least of my concerns.

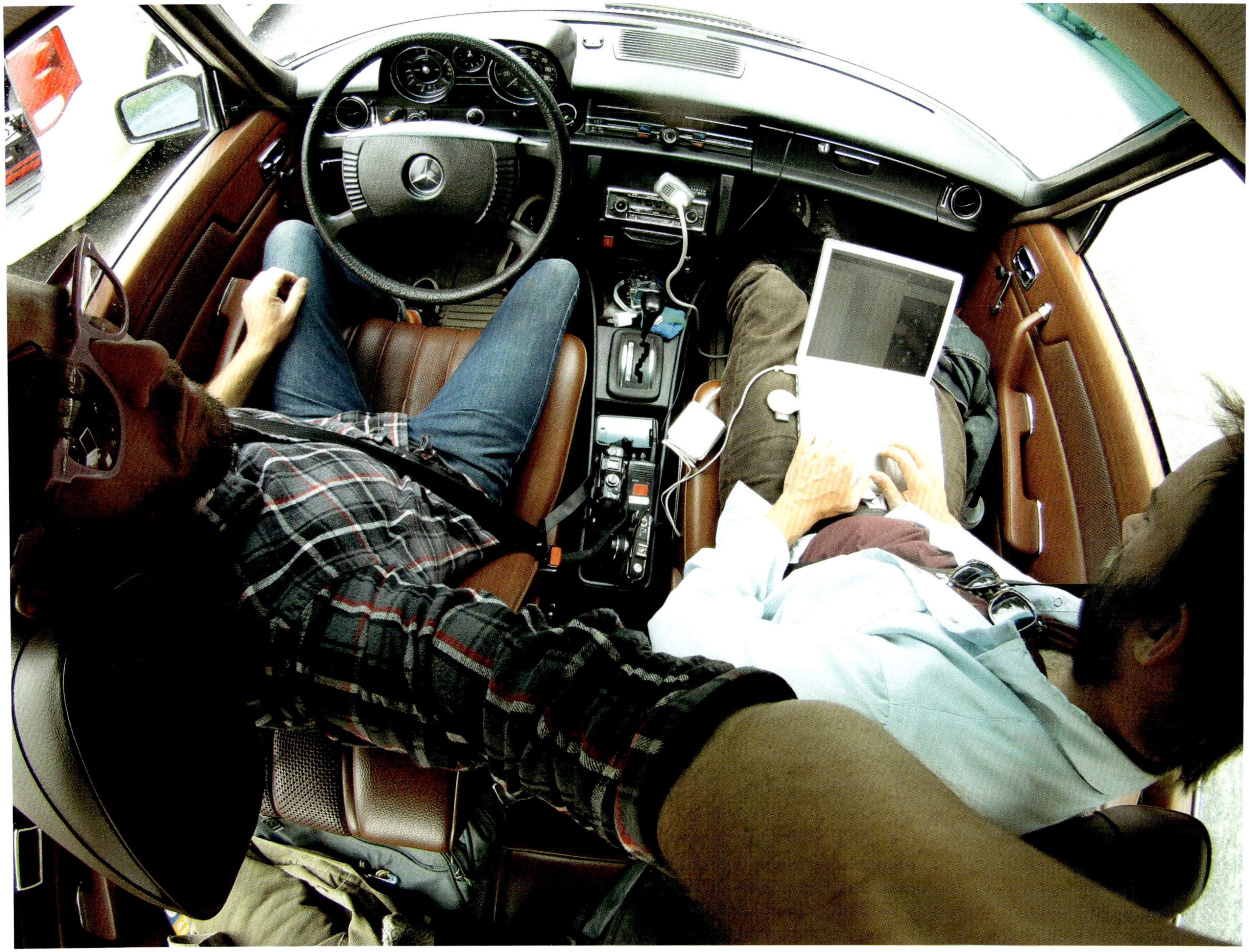

J.G. und Sean unterwegs in einem 300 Touring Diesel.
J.G. and Sean out and about in a 300 Touring Diesel.

[redacted]

Auf ihren Trips durch Amerika und darüber hinaus werden J.G. und Sean immer wieder Zeuge beeindruckender Naturschauspiele.
J.G. and Sean regularly witness impressive weather events on their trips through America and beyond.

Der Westen der USA besteht nicht nur aus unberührter Natur.
The US West is more than just untouched nature.

Geld haben werde, um all die Dinge zu tun, die ich in meinem Leben noch machen möchte, dann hätte ich wahrscheinlich gar nichts mit meinem Leben angefangen. Außerdem werde ich sowieso nie genug Zeit oder Geld haben, d.h. es lohnt sich gar nicht groß darüber nachzudenken. Wir haben hier auch gerade ganz andere Sorgen – zum Beispiel Elche, die über die Straße schlurfen oder Bären, die mein Bein für ein leckeres Sandwich halten könnten(...)Sean starrt aus dem Fenster. Sean starrt eigentlich immer aus dem Fenster. Dabei kommen ihm wahrscheinlich all seine wahnsinnigen Einfälle. Aber wer weiß das schon genau. Er hortet sie wie einer dieser verrückten Reality TV-Sammler. Irgendwann wird er die Welt mit so etwas wie dem Da Vinci Code erschüttern.
Wenn man bedenkt, dass wir beide unermüdlich durch diese Windschutzscheibe starren, könnte man eigentlich annehmen, dass wir uns keinen zweiten Ölwannenriss in unserer Roadtrip-Karriere leisten würden, aber weit gefehlt. Wir doch nicht. Wir sind viel zu sehr damit beschäftigt, alles in uns aufzusaugen, um so richtig auf die Straße zu achten. Noch nicht mal im Yukon, wo uns wirklich jeder auf die schlechten Straßenverhältnisse hingewiesen hatte. Also ist es wieder passiert – wir hatten plötzlich ein Loch in der Ölwanne, und natürlich lief dann auch noch alles in der Warteschlange vor der kanadischen Grenze aus. Der Regen verteilte das Öl über den schwarzen Teer und verwandelte die Straße um uns herum in einen schillernden Regenbogen. In echter Sean und J.G.-Manier haben wir dann noch einen draufgesetzt und das Auto aus der Schlange geschoben, während die Grenzpolizei schon auf uns zuhielt. Ein paar Worte nebenbei zu Grenzpolizisten: Sie sind absolut humorlos und unberechenbar. Sie konfiszierten erst mal unsere Pässe, grillten uns standrechtlich – doch wir bestanden die Prüfung mit Bestnoten. Dann riefen sie einen Abschleppwagen, der uns über die Grenze in eine Geisterstadt namens Beaver Creek brachte. Dort, an einem verregneten kanadischen Sonntagnachmittag, konnten wir

I have more important things to worry about – like a moose doing a little soft shoe across the road or a bear making a sandwich with my leg. (...) Sean is staring out the window. Sean is always staring out the window. I think this is where he develops all of his genius ideas, but honestly, I wouldn't know. He hoards them like one of those freaky reality shows. I just know one of these days he's going to drop some Da Vinci Code or E=mc2 on all of us. Since both of us are constantly staring through this windshield you would think we'd avoid tearing a hole in the oil pan for the second time in our road tripping career, but nope. Not us. No way. We're too busy taking it all in to pay attention to the road. Not even in the Yukon Territory, where we were told time and time again how bad the roads were going to be, not even there did we pay enough attention. And so it happened again – we tore a hole in the oil pan, and to make matters worse we dumped all the oil while sitting in line at Canadian Customs. Dispersed by the rain the oil spread across the wet blacktop and a rainbow appeared on the ground in every direction. In true Sean and J.G. form we outdid ourselves. We pushed the car out of line as the border patrol walked towards us. Let me tell you something about border patrol: they're serious and unpredictable. They took our passports, gave us the whole 20 questions bit and we aced the exam. They called a tow truck that pulled us across the border and into the one-horse-town of Beaver Creek. This is where on a cold

"YOU CAN PULL UP TO A CLUB WITH AN OLD 300 D AND PARK UP FRONT THE SAME AS IF YOU DRIVE A BRAND NEW SL. IT STILL HAS THIS STATUS FEEL."

Wie Perlen auf einer Kette aufgereiht warten J.G.'s und Seans Prachtstücke auf ihre Käufer.
Like a chain of pearls: J.G. and Sean's restored beauties wait for their buyers.

Sean and J.G. am Elpaso Flughafen in Texas.
Sean and J.G. at El Paso International Airport, Texas.

Jedes Jahr fahren die beiden tausende von Kilometern, immer auf der Suche nach dem nächsten Schatz.
The two friends drive thousands of kilometres every year, always on the hunt for their next treasure.

zwei ausgesprochen übellaunige alte Männer überreden, uns einen Imbusschlüssel, etwas Kaltschweißkleber, Bremsreiniger und einen Wagenheber zu leihen. Wir haben den Wagen dann am Straßenrand repariert, dort übernachtet und sind am nächsten Tag weitergefahren."

„WENN DU MIT EINEM ALTEN 300 D VOR EINEM ANGESAGTEN CLUB VORFÄHRST, HAT DAS EINE ÄHNLICHE WIRKUNG, ALS WENN DU DAS MIT EINEM NEUEN SL TUST. DER STATUS STIMMT!"

Doch zurück nach Los Angeles. J.G. sagt, Mercedes Motoring sei so etwas wie eine glückliche Fügung, die perfekte Lösung für einen wie ihn, der sich nie hätte vorstellen können, ein Leben lang hinterm Schreibtisch zu verbringen. Die Idee, seine Liebe für Classic Cars mit seiner Abenteuerleidenschaft zu verbinden, wäre der Kickstart gewesen. „Ich hatte alle möglichen Classic Cars", erzählt er. „Aber erst, als ich meinen ersten Mercedes-Benz kaufte, einen 1979er 300 SD, ausgekleidet mit ausgeblichenem gelben Ahorn und sonnengewohnten Bambus, wurde mir klar, dass man ein Vintage-Auto fahren kann, das tatsächlich absolut verlässlich ist. Ein Auto, mit dem du ganz Amerika erleben kannst, wie in diesen alten, lässigen Filmen. Ein Auto, mit dem du die Welt erobern kannst. Plötzlich hatte ich meine Aufgabe gefunden, etwas, das ich wirklich liebte."

Seine liebsten Projekte: „Mein 1969er 300 SEL 6.3 und mein 1979 250 T der Baureihe 123. Zwei Autos, die sehr unterschiedlich zu restaurieren waren, aber gleichermaßen beein-

and rainy Sunday afternoon in Canada we convinced two extremely grumpy old men to loan us an Allen wrench, some JB Weld, brake cleaner and a floor jack. We fixed the car roadside that afternoon, spent the night and drove on the next day. "

But back to Los Angeles. J.G. said Mercedes Motoring was something of a happy accident in his life – the perfect solution for a person who never imagined himself spending his life behind a desk. The inspiration for his work came from his love of classic cars combined with his passion for adventure.
"I had all kinds of classic cars", he says. "But it wasn't until I bought my first Mercedes-Benz, a 1979 300 SD painted in faded maple yellow and sunburned bamboo, that I realized you can drive vintage cars that are totally reliable. A car you can experience America with, just like in all those cool old movies. A car you can conquer the world with. All of a sudden I had found my purpose – something I really loved."

His favourite projects: "My 1969 300 SEL 6.3 and my 1979 250 T with an M 123. Those two were very different restoration projects but were both equally impressive when they were done. I needed a year for each."
His gift for making the best out of these magnificent vehicles is something he inherited from his mother.
"My mother raced cars in the sixties, she is very hands on and talented in fixing things so I guess I got it from her." As a child, he also had a neighbour with a Mercedes-Benz garage, an older man who became a good friend and had a major influence on J.G. as a child.
"The unmistakable sound of the engine, when he pulled out of the driveway, the way the door shut and the rich smell of the premium leather interior. It's in the details", he says about the Mercedes mystique. He is fascinated by the fact that "you can take the whole car apart and put it back together without any problems. That's when you have this moment of "wow, that's amazing."
And once they've found the right car, he and

Die Straßen Amerikas sind bereits ihr zweites Zuhause.
The roads of America are already their second home.

[redacted]

„EIN VINTAGE-MERCEDES IST EIN AUTO, MIT DEM DU GANZ AMERIKA ERLEBEN KANNST, WIE IN DIESEN ALTEN, LÄSSIGEN FILMEN. EIN AUTO, MIT DEM DU DIE WELT EROBERN KANNST.“

“A VINTAGE MERCEDES IS A CAR YOU CAN EXPERIENCE AMERICA WITH, JUST LIKE IN ALL THOSE COOL OLD MOVIES. A CAR YOU CAN CONQUER THE WORLD WITH.”

Fototapete oder Realität? Mercedes-Benz Youngtimer vor tropischer Kulisse.
Real, or a fake backdrop? A Mercedes-Benz modern classic in tropical surroundings.

Unterwegs im Westen von Texas.
On the road in western Texas.

In der Mercedes Motoring Garage gibt es alles, was es braucht, um die Mercedes-Benz Klassiker originalgetreu wiederzubeleben. Ein Ort nicht nur für Technik, sondern auch für Geist, Seele und Geschichte.
The Mercedes Motoring garage has everything needed to restore a classic Mercedes-Benz to original condition. It's not just a technical space, but an emotional and historical one too.

[redacted]

druckend nach ihrer Fertigstellung, die jeweils ein ganzes Jahr dauerte. Das Talent, das Beste aus großartigen Fahrzeugen herauszuholen, hat J.G. sicher von seiner Mutter. „Sie fuhr Rennen in den Sechzigern und hatte immer eine Gabe, beim Autobasteln Dinge hinzukriegen.“ Und dann war da noch sein Nachbar mit der Mercedes-Benz Werkstatt, der ein guter, wenn auch älterer Freund wurde und in dessen Garage J.G. einen spannenden Teil seiner Kindheit verbrachte.
„Der unverwechselbare Sound der Motoren, wenn er einen Wagen die Ausfahrt hinunterfuhr, das satte Schnappen, mit dem sich Türen schlossen, der würzige Ledergeruch des hochwertigen Interieurs. Das sind alles Details, die ein Fahrzeug mit dem Stern für mich ausmachen. Bis hin zu der faszinierenden Tatsache, dass man den ganzen Wagen komplett auseinandernehmen kann und dann wieder zusammensetzen bis zu diesem Moment der Perfektion, der einfach glücklich macht.“ Und dann raus auf die Straße. Nicht auf die Interstate, nein auf die kleinen, abgelegenen Straßen. „Meist übernachteten wir unterwegs bei Freunden, schlafen in Hostels – oder – eine meiner Lieblingserinnerungen – einfach am Strand und hören den Kokosnüssen zu, wie sie in den weichen Sand plumpsen“, erinnert sich Sean.
Mercedes Benz Klassiker – für J.G. und Sean ist das der perfekte Lifestyle. „In L.A. gibt es immer mehr junge Leute, die die Individualität zu schätzen wissen, die ein Vintage-Mercedes mit sich bringt“, weiß J.G. „Nicht jeder kann sich ein neues Auto leisten. Aber wenn du mit einem alten 300 D vor einem angesagten Club vorfährst, hat das eine ähnliche Wirkung, als wenn du das mit einem neuen SL tust. Der Status stimmt!“
Wir verraten jetzt mal, dass der 300 D von 1975 braun mit lavendelfarbenen Streifen lackiert ist, die perfekt zu J.G.'s Ray Ban passen. Cooler lässt sich ein Auto wohl kaum personalisieren.
Wer sich davon persönlich überzeugen möchte, muss sich allerdings weiter gedulden. Es ist, wie eingangs erwähnt, nicht leicht, J.G. zu treffen. Das nächste große Abenteuer ist bereits in Arbeit: Gemeinsam mit Sean will er einen der sehr seltenen Mercedes-Benz W 115 Kombi in Argentinien kaufen und ihn im Frühling 5.000 Kilometer durch sieben südamerikanische Länder fahren. „Da sind wir für die Autosuche schon jetzt viel unterwegs und auch, um die genaue Route zu planen.“
Die wird ganz sicher wieder spektakulär. Man darf sich also auf neue spannende Reiseberichte und tolle Vintage-Mercedes Fotos auf mercedesmotoring.com freuen. •

Sean head out. Not on the interstate: they take the small side roads. “We usually stay overnight with friends, occasionally sleep in hostels or – one of my favourite memories – crash on the beach and listen to coconuts falling to the ground”, Sean recalls.

Mercedes-Benz classics are the perfect lifestyle for J.G. and Sean. “In Los Angeles there are many young people who appreciate the kind of individuality an older car brings. Not everyone can afford a brand-new car but classic cars still have this recognition. You can pull up to a club with an old 300D and park up front the same as if you drive a brand new SL. It still has this status feel”, J.G. says.
We're going to go ahead and reveal that the 1975 two-toned brown 300 D with lavender-coloured details is a perfect match for J.G.'s Ray Bans. There could not possibly be a cooler customised car...
If you want to see for yourself, though, you will have to keep waiting for a while. As we said at the beginning, it's not easy to meet up with J.G. His next big adventure is being planned: he and Sean want to buy some of the very rare Mercedes-Benz W 115 station wagons in Argentina and drive them 5000 kilometres through seven South American countries in the spring. “We are already on the road a lot as we look for the cars and plan the exact itinerary.”
No doubt it will be another unforgettable trip. We can look forward to exciting new updates and pictures of fantastic vintage Mercedes at mercedesmotoring.com. •

Der Weg ist das Ziel: Neben der Restauration gehören immer wieder auch Road Trips zum Erfahrungsschatz des Mercedes Motoring Teams.
The journey is the destination: the Mercedes Motoring team are experienced in road trips as well as restoration.

J.G. und Sean auf einem ihrer Trips in den wilden Südwesten der USA.
J.G. and Sean on one of their trips in the rural southwestern US.

Ihr Begleiter: Ein zweifarbiger Mercedes-Benz 300 D der Baureihe W 115.
Their companion: a two-colour Mercedes-Benz 300 D (W 115).

432.7 km/h

DER SCHNELLSTE PFEIL DER WELT
THE WORLD'S FASTEST "ARROW"

ÜBER SEINE REKORDFAHRT 1938 SAGTE RUDOLF CARACCIOLA: „BEI DER GESCHWINDIGKEIT, DIE ICH DURCHFUHR, MUSSTE ICH SEHR EXAKT LENKEN. DOCH NOCH BEVOR DAS GEHIRN DIE NOTWENDIGE HANDLUNG ERKANNT HATTE, WAR DER WAGEN BEREITS VORBEIGERAST."

RUDOLF CARACCIOLA ON HIS RECORD-BREAKING DRIVE IN 1938: „AT THE SPEED I WAS TRAVELLING AT, I HAD TO STEER EXTREMELY PRECISELY. BUT BEFORE MY BRAIN HAD FIGURED OUT WHAT TO DO, THE CAR HAD ALREADY RACED FURTHER AHEAD."

(...) Viele sagten über uns: „Das ist verrückt, was die wagen!“
So beschrieb Rennfahrerlegende Rudolf Caracciola, was er und seine Rennfahrerkollegen in den 1930er Jahren hinter dem Lenkrad ihrer Silberpfeile vollbrachten.
„(...) A lot of people said ‘That’s crazy, what they’re attempting!’“
recalled motorsport legend Rudolf Caracciola once, referring to his and his colleagues‘ endeavours behind the wheels of their Silver Arrows.

„320 Stundenkilometer – viele sagten über uns: „Das ist verrückt, was die wagen!“ Rennfahrerlegende Rudolf Caracciola beschrieb selbst recht milde das, was er und seine Rennfahrerkollegen in den 1930er Jahren unter harter Arbeit hinter dem Lenkrad ihrer Silberpfeile vollbrachten: Rekorde über Rekorde. Aber das war erst der Anfang. In aller Munde ist bis heute sein ungebrochener Geschwindigkeitsrekord aus dem Januar 1938 im legendären Mercedes-Benz W 125 12-Zylinder-Rekordwagen, mit dem er sich in der Automobilgeschichte verewigte: 432,7 Kilometer pro Stunde auf einer öffentlichen Straße.

Um zu verstehen, wie es zu diesem Rekord kam, muss man die Zeit verstehen. Anfang der 1930er Jahre war das Automobil kurz davor, eine neue Evolutionsstufe zu erreichen. Stärkere Motoren und neuartige Chassis-Konstruktionen ermöglichten eine größere Leistungsausbeute. Da scheint es nicht verwunderlich, dass die Jagd nach Leistung und Geschwindigkeit in diesen Jahren so gut wie alles im Motorsport bedeutete, gefühlt mehr noch als der eigentliche Zweikampf zwischen Mann und Mann. Und den trugen die Piloten nicht nur gegen die Konkurrenz aus, sondern auch gegen sich und vor allem gegen die Zeit. Über- dem italienischen Namen und Wurzeln, der in Italien bei seinen Erfolgen als erfolgreicher Ausländer gefeiert wurde, bestimmten bei Mercedes-Benz in diesem Jahrzehnt das Renngeschehen auf dem Asphalt. „Caratsch“, wie man ihn schon fast liebevoll überall in Szene und Presse nannte, war sozusagen die eigentliche Gallionsfigur der Stuttgarter Marke mit dem Stern. Zumindest das sportliche Aushängeschild, denn seine Karriere als Rennpilot begann in den 1920er Jahren und er startete sehr schnell durch, legte im Rennwagen W 25 von 1934 erste Rekordfahrten hin, wurde 1935 sogar Europameister. Es folgten 1937 ein zweiter Europameister-Titel und zahlreiche Siege im legendären W 125, der bis zu 646 PS/475 kW aus acht Zylindern zauberte und eine Höchstgeschwindigkeit von über 320 km/h ermöglichte. Für damalige Verhältnisse – und natürlich auch noch für heutige – ein ungeheuer hohes Tempo, das einem Rennfahrer nicht nur einiges an Reaktion und Konzentration, sondern auch Muskelkraft abverlangte. Angesichts der heute mit allen technologischen Raffinessen ausgestatteten Rennboliden eine unvorstellbare Aufgabe. Doch die Bindung Mann und Maschine bedeutete in diesem Jahrzehnt sehr viel, galt Automobiltechnologie doch in Zeiten

UM ZU VERSTEHEN, WIE ES ZU DIESEM REKORD KAM, MUSS MAN DIE ZEIT VERSTEHEN.

haupt war es wohl auch die Zeit, in der der Begriff „einen heißen Reifen fahren“ entstanden sein muss, denn ein Hochgeschwindigkeitsrennen jagte das andere.
Manfred von Brauchitsch, Hermann Lang und allen voran Rudolf Caracciola, der Mann mit des weltweiten Börsencrashs auch als wirtschaftlicher Wegweiser. Das spornte an. Daimler-Benz schuf schließlich den Mercedes-Benz W 125 Rekordwagen, mit dem inzwischen zum Cheftestfahrer aufgestiegenen Rudolf Caracciola 1938 den unglaublichen

“320 kilometres per hour – at the time, a lot of people said ‘That’s crazy, what they’re attempting!’” recalled motorsport legend Rudolf Caracciola once. “Crazy” is an understatement when it comes to the records that he and his teammates worked so hard on in the 1930s in the cockpits of their Silver Arrows. The speed record that Caracciola set in January of the motorsport side of the brand. His racing career began in the 1920s and he quickly reached the top of the sport. Caracciola’s first record-setting drives were in the W 25 racing car in 1934. The following year, he won the European Drivers’ Championship, going on to win it again 1937. From that year on, he notched up a string of victories in the legendary

TO UNDERSTAND HOW THIS WORLD RECORD WAS SET, YOU HAVE TO UNDERSTAND THE TIMES.

1938 in his legendary 12-cylinder Mercedes-Benz W 125 Rekordwagen – 432.7 kilometres per hour on a public road – is still unbroken today, ensuring him and the car a lasting place in automotive history.

To understand how this world record was set, you have to understand the times. At the start of the 1930s, the automobile was about to enter a new stage of its evolution. More powerful engines and brand new chassis designs made higher performance possible. The race for power and speed was all-important in motorsport – more so than the competition between drivers. Racing drivers weren’t just competing against each other, they were competing against themselves and most of all against the clock. It was the era of hot rubber, when one high-speed race followed the next.

Racing drivers Manfred von Brauchitsch, Hermann Lang, and most of all Rudolf Caracciola – an Italian-German especially celebrated in Italy – dominated the decade at Mercedes-Benz. Caracciola, or “Caratsch”, as the German media and fans called him, was almost a figurehead for Mercedes-Benz, a personification W 125, an 8-cylinder racing car with a maximum power of 637 hp (482 kW) that was able to reach over 320 km/h: incredible for the time and still remarkable now. Driving at this velocity demands real skill – not just rapid reaction and complete concentration but also strength and muscularity. Modern racing cars designed to be driven at speeds this high boast all kinds of technological refinements, so for a present-day racing driver, driving at 320 km/h in such a basic car would be inconceivable. The 1930s was a decade when the relationship between man and machine meant a lot to people – during the Great Depression, automotive technology was seen as the economic future, an encouraging ray of hope on the horizon.
Daimler-Benz purpose-built the W 125 Rekordwagen in which Caracciola, by now chief test driver, set his phenomenal speed record in 1938. Its approach to aerodynamics was revolutionary. Taking the successful W 125 Silver Arrow as a starting point, the engineers created an ultra-low, streamlined body with two large air inlet nozzles at the front and a protruding, wedge-shaped rear. Wind tunnel tests showed the full potential of this novel body

Der Mercedes-Benz 12-Zylinder-Rekordwagen W 125 fuhr die schnellste je auf einer öffentlichen Straße erreichte Geschwindigkeit.
The 12-cylinder Mercedes-Benz W 125 Rekordwagen achieved the highest land speed ever on a public road.

Geschwindigkeitsrekord aufstellen sollte, nach völlig neuen aerodynamischen Gesichtspunkten. Den erfolgreichen Silberpfeil W 125 als Basis, konstruierten die Ingenieure eine völlig neue, sehr flache, stromlinienförmige Karosserie, die an der Front zwei große Lufteinlassstutzen hatte. Das Heck hingegen schneiderten sie ausladend und keilförmig. Durch Tests im Windkanal zeigte sich dann das ganze Potential dieser neuartigen Karosseriestruktur. Die Techniker erzielten einen cW-Wert von 0,157, sensationell für die damalige Zeit – und auch noch für heute! Der Silberpfeil nun also in neuem Gewand. Um den künftigen Rekordwagen einsatzfähig zu machen, verpasste man ihm als Herzstück einen großen V 12 Kompressormotor, dessen Leistung sich durch zwei zusätzliche Vergaser von 570 PS (419 kW) auf über 730 PS (536 kW) steigern ließ. Gekühlt wurde er bei Last mit 500 Liter Eis und Wasser.

Dann, am 28. Januar 1938 war es so weit, eine der spektakulärsten Rekordfahrten der Geschichte stand kurz bevor. Da es keine Rennpisten gab, die lang genug für solche Rekordfahrten waren, hatte man sich für die Autobahn entschieden, denn die meist sehr langen und kerzengeraden Autobahnen in Deutschland boten sich perfekt für Hochgeschwindigkeits-Tests an. So sperrte man für diese Testfahrt den Autobahnabschnitt zwischen Frankfurt am Main und Darmstadt komplett ab. Carracciola, am Steuer des W 125 Rekordwagens, war wohl sicherlich innerlich angespannt aber ebenso erwartungsvoll und präsentierte sich gewohnt gelassen und souverän. Mit einer wahnsinnigen Beschleunigung katapultierte er den W 125 auf einer Distanz von einem Kilometer auf 432,7 km/h bei fliegendem Start. Caracciola hatte es geschafft. Weltrekord. Ungebrochen bis heute. •

structure: the technicians achieved a drag coefficient of 0.157, sensational for the time and still amazing today. The Silver Arrow had metamorphosed. To ready the car for the record attempt, the team fitted it with a large 12-cylinder supercharged engine with a maximum power of 562 hp (419 kW), aided by two additional carburettors which brought the vehicle's total power to over 720 hp (536 kW). The car was cooled using 500 litres of ice and water.
28 January 1938 was the date for one of history's most spectacular land speed record attempts. As there was no racetrack suitable, it was decided to use the autobahn – long, straight and thus perfect for driving at extreme speed. The autobahn was closed between Frankfurt and Darmstadt for the attempt.
Caracciola, at the wheel of the W 125, undoubtedly somewhat nervous on the inside, appeared typically relaxed, confident and positive. With incredible acceleration, he catapulted the W 125 from 0 to 432.7 km/h in just one kilometre. He'd done it – he'd set a new world record. One that's still unbroken today, 75 years on. •

Rudolf Caracciola ein Könner am Steuer
des legendären W 125 Silberpfeils.
Rudolf Caracciola: an expert behind the
wheel of his legendary W 125 Silver Arrow.

Sie schrieben Motorsportgeschichte (von li. nach re.): Mercedes-Benz 1,5-l-Rennwagen W 165 (1939), Mercedes-Benz Rekordwagen T 80 (1939), Mercedes-Benz 12-Zylinder-Rekordwagen W 125 (1938).
They made motorsport history (from left to right): the 1.5-litre Mercedes-Benz W 165 Rennwagen (1939), the Mercedes-Benz T 80 Rekordwagen (1939) and the 12-cylinder Mercedes-Benz W 125 Rekordwagen (1938).

Das Rennplakat des Weltrekords: Caracciola fährt am 28. Januar 1938 zu neuen Rekorden.
Caracciola's new world records were communicated to the public by poster.

Rudolf „Caratsch“ Caracciola mit seinem Hund Moritz.
Rudolf „Caratsch“ Caracciola with his dog Moritz.

[redacted]

78
80
82
84
Borgo
269
2012
S OK 54 H

MILLE MIGLIA

DIE LEGENDÄREN TAUSEND MEILEN
THE LEGENDARY 1000 MILES

BRAVISSIMA, FANTASTICA, STUPENDA - IN DER ABENDDÄMMERUNG MISCHEN SICH ADJEKTIVISCHES FACHSIMPELN UND ENTHUSIASTISCHE AUSBRÜCHE IN SCHWINDELERREGENDEM ITALIENISCH-STAKKATO MIT RENNUMMERN, FAHRERNAMEN UND MODELLBEZEICHNUNGEN. WILLKOMMEN BEI DER MILLE MIGLIA.

BRAVISSIMA, FANTASTICA, STUPENDA – CAR DESCRIPTIONS, RACING NUMBERS, AND DRIVER NAMES EXPLODE OUT INTO THE DAWNING NIGHT IN RAPID-FIRE ITALIAN STACCATO. WELCOME TO THE MILLE MIGLIA.

Rudolf Caracciola und Wilhelm Sebastian beim Start zur Mille Miglia 1931 in Brescia.
Rudolf Caracciola and Wilhelm Sebastian at the start of the 1931 Mille Miglia in Brescia.

[redacted]

Das Spektakel füllt auch heute noch die Straßen und Plätze italienischer Städte wie Bologna, Rom und Ferrara.
Even today, the spectacle fills the streets and squares of Italian cities like Bologna, Rome and Ferrara.

Die teilnehmenden Oldtimer finden sich auf dem Gelände des Museo Mille Miglia ein, bevor um 18.30 Uhr der Startschuss fällt. Durch das Tor und Verona direkt nach Ferrara, dem ersten offiziellen Halt des Rennens. Unterwegs, am Ufer des Gardasees, locken flüchtig Terassentische mit ihren leuchtend-orangenen Gläsern voll Aperol im warmen Abendlicht. Doch schon bald dominiert der Sound von Motor und Auspuff die Idylle, als ein Mercedes-Benz SSK, ein Alpha Romeo 6C 1500 und ein Bugatti A35 über das Kopfsteinpflaster brettern. Die versammelten Zuschauer jubeln den Fahrern und ihren Lieb-

VOLLGAS VORAUS - ES GEHT VON BRESCIA NACH ROM.

lingen unermüdlich zu; ihre Kinder lehnen sich wagemutig über die Absperrung, um die Namen ihrer Favoriten hinauszukrähen, die dankbar zurückwinken.

Am folgenden Tag geht es aufwärts, genauer gesagt 700 Höhenmeter nach San Marino, durch die engen Gassen der Stadt bis zur mittelalterlichen Piazza des Palazzo Pubblico. Ohne seine Augen auch nur eine Sekunde von der glorreichen Prozession abzuwenden, freut sich der Barista des Cafe Piadineria: „Das ist alles Teil der Geschichte, all die Wagen hier auf der Straße." Nach kurzer Pause setzt er nach, „es ist nett." Angesichts der fast hypnotischen Aufmerksamkeit, die er dem Spektakel widmet, halten wir seine persönliche Definition von „nett" für dezent untertrieben.

Die Strecke führt weiter nach Sansepolcro und schließlich nach Rom. Paolo, der einen der legendären 300 SLS-Flügeltürer fährt, genießt „das unvergessliche Erlebnis, mit

The classic beauties rumble into the grounds of the Museo Mille Miglia before bursting out of the gates at 6:30 p.m. to speed through Verona and into Ferrara, the scheduled first stop of the race. On their way, they pass along-

PEDALS TO THE FLOOR IN BRESCIA, SIGHTS SET ON ROME.

side Lago di Garda where tables are dotted with bright orange glasses of Aperol, glinting in the sunset. Before long, the music of engines and exhaust pipes resonates over the lake and the likes of the Mercedes-Benz SSK, Alfa Romeo 6C 1500 and Bugatti A35 thunder into view over the cobblestones. The crowd doesn't seem to tire of the cheering, and children lean over guardrails, screaming the names of their favourite drivers, who wave back in appreciation.

The following day, the cars climb 700 metres to San Marino, exploding between narrow streets and into the piazza fronting the Palazzo Pubblico. Without taking his eyes off the line-up of cars, the barista at Café Piadineria remarks: "This is part of history, to see all these cars on the road." The cars then descend down into Sansepolcro on to Rome. Paolo, driving one of the legendary 300 SL Gullwings, says it is "an unforgettable experience to drive this wonderful car through the paesaggi of Italia." Frank Lamparter of the SLR Club is of a similar view, and eagerly anticipates the opportunity to "sight-see in Rome in an SLR".

Once the cars have arrived in front of the flood lamps of Castel Sant'Angelo, the race announcer encourages one driver to "pump up the engine a bit". In turn, the engine roars, and the announcer shouts "La muuuuusica!" into the mic. Mick Walsh, the editor-in-chief

Eine gute Mischung: Historische Altstädte und Mercedes-Benz Klassiker wie dieser 300 SL Rennsportwagen (W 194).
The perfect mix: historic city centres and classic Mercedes-Benz automobiles like this 300 SL racing car (W 194).

[redacted]

diesem wundervollen Auto durch die italienischen paesaggi zu brausen." Und auch Frank Lamparter vom SLR Club freut sich sehr auf die Chance, „die römischen Sehenswürdigkeiten im SLR abzuklappern."

Im Flutlicht vor dem Castel Sant'Angelo animiert der Moderator einen Fahrer, „den Motor mal ordentlich röhren zu lassen", und ruft beim folgenden imposanten Krach enthusiastisch „La muuuuusica!" ins Mikrofon. Und selbst Mick Walsh (Chefredakteur des Classic & Sports Car-Magazins), der in Brescia noch leichte Bedenken wegen der versammelten Pressemeute hatte, lässt sich in Rom im schnittigen 300 SLR (W 196 S) mit leichtem Grinsen zu einer Äußerung hinreißen, die die Veranstaltung treffend auf den Punkt bringt: „Was für eine verrückte Idee, mit all diesen Wagen in Roms Zentrum einzufallen. Es ist ein echter Traum!"

Aber machen wir uns nichts vor: Die Mille Miglia ist keine gemütliche Sonntagsspazierfahrt durch Malibu, sondern ein unglaublich anstrengendes Rennen, das sämtlichen Fahrern, Mechanikern, Pressevertretern, Helfern, und Organisatoren drei Tage lang hundert Prozent abverlangt. Und obwohl an Schlaf kaum zu denken ist, buhlen Fahrer aus allen Ecken und Enden der Welt um eine der begehrten Startnummern für dieses berühmt-berüchtigte Rennen. Auf der Strecke präsentieren sie sich folglich ebenso elegant und einzigartig wie ihre liebevoll gehegten Fahrzeuge: Die Fahrer erscheinen in Leinen

DAS ELEGANTE ERBE DES MERCEDES-BENZ-FAHRGEFÜHLS – DIE TOSKANISCHE LANDSCHAFT IM TAKT DER MOTOREN.

of Classic & Sports Car Magazine, expressed some anxiousness in Brescia due to the onslaught of press photographers, but by this point in Rome, he's content. Walsh and Mass are seated in their sleek 300 SLR (W 196 S), and Walsh's smile says it all. "What a crazy idea to bring all these cars into the centre of Rome", he says. "It's a lovely dream!"
Make no mistake: the Mille Miglia is not a Sunday drive through the Malibu canyons. An incredibly arduous race, the schedule calls for drivers, mechanics, press, administration and all involved to run through the nights with hardly any sleep. Despite this, drivers are here from all corners of the world to take part in this famous event, and they look just as elegant and diverse as their cars: linen and silk scarves, leather driving caps, racing goggles, chequered shirts, flowers through tailored lapels, tartan arm patches, chronograph watches, and driving gloves.
With a 4:30 am rise, the third day is the longest and hardest stretch of the Mille Miglia, from Rome all the way back to Brescia. Over the course of the morning, the mist slowly dissipates and golden light escapes from between the blue-black silhouettes of the Tuscan mountains. The early-20th century automobiles kicking up dust in this ancient landscape, racing through yellow-stoned cities, make for a surreal yet oddly appropriate sight. The muted colours of the car bodies blend beguilingly with the rustic backdrop.
Outside of Florence, a gentleman dressed in sporty formals is violently waving a white handkerchief from his roadside table at

MERCEDES-BENZ HERITAGE: AN EXHILARATING DRIVING EXPERIENCE THROUGH TUSCANY.

AVTOMOBILE CLUB
BRESCIA
DUNLOP

Zielduchfahrt in Brescia für das Siegerteam Caracciola / Sebastian. Damit gewinnt zum ersten Mal eine ausländische Fahrerpaarung.
Caracciola and Sebastian pass the finish line in Brescia. Their victory was the first by a non-Italian duo.

Die Mille Miglia ist ein unglaublich anstrengendes Rennen. Dennoch fährt Stirling Moss 1955 mit seinem 300 SLR zu einem Rekord für die Ewigkeit.
The Mille Miglia is an incredibly tough race. Despite this, in 1955, Stirling Moss managed to set a new record time in his 300 SLR that remains unbroken.

BRESCIA REPRÄSENTIERT DEN MERCEDES-BENZ-EIGENEN RENNSPORTSGEIST BEI DER MILLE MIGLIA.

und Seidentüchern, ledernen Schutzhelmen, Schutzbrillen, karierten Hemden, Blumen im maßgeschneiderten Revers, Schottenkaro-Ellenbogenflicken, Einsteckuhren und sportlichen Autohandschuhen – ein wahrhaft zeit loser Look.

Am dritten Tag wartet der längste und härteste Abschnitt der Mille Miglia auf alle Versammelten – von Rom bis zurück nach Brescia. Pünktlich zum Weckruf um 4:30 Uhr verzieht sich der Morgennebel und goldenes Licht umspielt die von blauschwarzen Bergsilhouetten eingerahmten toskanischen Täler. Fast schon surreal mutet es an, als wenig später die glorreichen Produkte des frühen 20. Jahrhunderts viel Staub in dieser altertümlichen Landschaft mit ihren gelb geziegelten Städten aufwirbeln. Gemeinsam mit drei Mercedes-Benz-Klassikern, einem Ferrari und einem Bugatti nehmen wir die Biegungen der geschwungenen Straßen nördlich von Siena. Das sanfte Rot, gleißende Schwarz, dunkle Grün und gedämpfte Weiß der Oldtimer-Karossen verschmelzen mit der bllühenden Landschaft zu einem einzigartigen Bild. Kurz vor Florenz winkt uns ein echter Gentleman in sportlich-formeller Kleidung mit einem weißen Taschentuch an seinen Tisch draußen vor der Trattoria con Giardino. Sein stolzer Einwurf: „Ich kann mich noch an die echte Mille Miglia erinnern! Da waren Sie noch nicht geboren, aber vor fünfzig Jahren habe ich das echte Rennen mit eigenen Augen gesehen. Insgesammt 1.500 Autos... und viele Menschen, die den Wagen hinterhergerannt sind!" Aber bei der Mille Miglia herrscht nicht

BRESCIA EMBODIES THE MERCEDES-BENZ RACING SPIRIT AT THE MILLE MIGLIA.

Trattoria con Giardino. He proudly recounts: "I remember the real Mille Miglia! Fifty years ago, I was at the real races – 1,500 cars in total, with people running after the cars!" Then and now, the Mille Miglia has never been without challenges. Italy's weather can be as boisterous and tempestuous as its people; the drivers must contend with sudden hard rain as they race on treacherous mountain roads. Undeterred, they pull the collars of their outer coats up against the oncoming rain and accelerate their way towards the finish line.

In Brescia, the sidelines on Viale Venezia are already packed long before the cars arrive – and everyone has a story to tell. The team in the Mercedes-Benz Pan Americana SL boasted that their car experienced "no technical problems" and had "more capacity than expected in terms of speed and driving precision." Most of these cars are works of art just as much as they are works of precision. North American Tribute driver Sylvia Oberti, driving solo, describes the Mille Miglia as "an endurance marathon. My car is limping to the finish line." The true racing spirit of this event comes to light as some cars, without barely enough strength left to make the last several hundred metres, receive welcome assistance from their competitors, who push them forward to the finish flag.

Setting aside the bloodshot eyes, pale, sleep-deprived faces, and hunched shoulders, every single driver wears a happy grin as they look at the crowds and each other, recognising the significance of this historical and

Obwohl an Schlaf kaum zu denken ist, buhlen Fahrer aus allen Ecken und Enden der Welt um eine der begehrten Startnummern.
Although sleeping during the Mille Miglia is almost unthinkable, drivers from every corner of the world clamour for a place in the race.

nur eitel Sonnenschein. Ebenso wie das italienische Volk schlägt das regionale Wetter gern Kapriolen und zeigt sich abwechselnd von seiner schönsten und stürmischsten Seite – ein plötzlicher Regenguss auf tückischer Bergstrecke ist hier keine Seltenheit. Doch davon lässt sich kein Teilnehmer abschrecken: Unbeirrt wird einfach der Mantelkragen hochgeschlagen und weiter gen Ziel gerast.

In Brescia sind die Bürgersteige der Viale Venezia schon dicht gesäumt, lange bevor die ersten Wagen eintreffen – und ihre Fahrer aus dem Nähkästchen plaudern. Nicht nur das Mercedes-Benz-Pan-Americana-SL-Team freute sich, dass es „keine technischen Probleme und deutlich mehr Spielraum als erwartert in Sachen Tempo und präzises Handling“ gab. Kein Wunder, denn viele dieser Fahrzeuge sind gleichzeitig Wunderwerke präziser Technik und absolute Kunstwerke. Andere, wie die US-amerikanische Tribute-Fahrerin Sylvia Oberti, die die Strecke allein zurücklegte, beschreiben die Mille Miglia als „einen wahren Ausdauermarathon. Mein Auto ist quasi über die Ziellinie gehumpelt.“ Und der wahre Sportsgeist der Veranstaltung zeigt sich, als einige Autos, denen schlicht die Kraft für die letzten paar Hundert Meter fehlt, bereitwillig von ihren Konkurrenten bis ins Ziel geschoben werden.
Wenn man einmal von den blutunterlaufenen Augen, blass-müden Gesichtern und hochgezogenen/steifen Schultern absieht, dann steht jedem der teilnehmenden Fahrer angesichts der Bedeutung dieses historischen, ehrenhaften Rennens ein glückliches Grinsen ins Gesicht geschrieben. Regine Sixt im Mercedes-Benz 300 SL dazu: „Es ist so aufregend, dass ich nicht die geringste Müdigkeit spüre!“ Einige Fahrer springen aus ihren Autos, ums sich gegenseitig zu gratulieren, ein herzliches Schulterklopfen entgegenzunehmen oder für Fotografen zu posieren. Selbst in Mailand, am Folgetag, lässt sich eine Dame beim Anblick des Pressepasses mit historischem Rennlogo zu einem aufgeregten „Ah! La Mille Miglia, la corsa più bella del mondo. Molto molto bella!“ hinreißen.

Und Recht hat sie/Und sie liegt goldrichtig. Schließlich haben all diese Autos eine Seele und eine eigene Stimme. Und wo ließe sich besser singen als bei der Mille Miglia? •

honourable race. Regine Sixt, in a Mercedes-Benz 300 SL, says: “It is so exciting, I am not even tired!” Some drivers jump out of their cars to slap each other on the back, exchange congratulations and pose for photographers. Even in Milan the following day, a lady sights the press pass emblazoned with the historic race logo, and excitedly remarks: “Ah! La Mille Miglia, la corsa più bella del mondo. Molto molto bella!” (“The most beautiful race in the world! Very very nice!”).

And she’s spot on. After all, these cars have spirit and voice. So where better for them to sing out in joy than at the Mille Miglia? •

Nicht nur das Zuschauerinteresse ist groß, auch für viele Medien ist das historische Event ein Highlight.
The historic event is a highlight for the media as well as spectators.

[redacted]

DAS MUSEO DIE SANTA GIULIA IN BRESCIA ZEIGT MIT DER AUSSTELLUNG „NOVECENTO MAI VISTO. FROM ALBERS TO WARHOL TO (NOW)“ DIE HIGHLIGHTS AUS DER DAIMLER KUNSTSAMMLUNG.

HIGHLIGHTS FROM THE DAIMLER ART COLLECTION WERE SHOWCASED IN THE EXHIBITION “NOVECENTO MAI VISTO. FROM ALBERS TO WARHOL TO (NOW)” AT MUSEO DI SANTA GIULIA IN BRESCIA, ITALY.

INSPIRATION AM STRECKENRAND

TRACKSIDE INSPIRATION

Vorherige Seite:
Andy Warhol
Mercedes-Benz 300 SL Coupé, 1954, 1986, Siebdruck, Acryl auf Leinwand, 152,5 x 127,5 cm

Previous page:
Andy Warhol
Mercedes-Benz 300 SL Coupé, 1954, 1986, screen print, acrylic on canvas, 152.5 x 127.5 cm

Albert Hien
Gib Gas, 1999
Glas, Argongas, Neongas
140 x 230 x 15 cm

Albert Hien
Step On The Gas (Gib Gas), 1999
Glass, argon, neon
140 x 230 x 15 cm

Auch mit dem Museo Mille Miglia in Brescia besteht eine Kooperation der Daimler AG, bei der in diesem Jahr historische Rennwagen kunstvoll in Szene gesetzt wurden. Bilder der Serie „Cars" von Andy Warhol, die in den 80er Jahren bei dem Künstler in Auftrag gegeben wurde, waren ausgestellt.

As part of its collaboration with Daimler, Museo Mille Miglia in Brescia is showcasing historic racing cars in front of artistic backdrops. Pictures from Andy Warhol's "Cars" series from the 1980s are also being exhibited.

Verbunden durch eine gemeinsame Leidenschaft für den historischen Motorsport und einer Begeisterung für die Kunst, fühlt sich Mercedes-Benz als bedeutender Teil des Faszinosums Mille Miglia, der Region besonders verpflichtet. Eine Kooperation zwischen der Daimler AG, dem Museo di Santa Giulia und dem Museo Mille Miglia in Brescia soll diese Bindung weiter stärken. Ziel ist es, die bedeutsame Verknüpfung von Mercedes-Benz Classic mit der norditalienischen Region und der Mille Miglia als wohl berühmtestes Straßenrennen Europas mit großer Historie, zu hervorzuheben und zu vertiefen.

So ist Mercedes-Benz die einzige ausländische Automobilmarke, welche die „1000 Meilen von Italien" auf der Originalroute zwei Mal gewinnen konnte. 1931 durch Rudolf Caracciola und Wilhelm Sebastian auf einem Mercedes-Benz SSK, 1955 durch einen Sieg von Stirling Moss und Denis Jenkinson mit einer Bestzeit, die bis heute ungebrochen ist. Die Kooperation mit den Museen würdigt die Renngeschichte – geht aber weit darüber hinaus.

Das, wie auch die Patentschrift für den Motorwagen von Carl Benz, 2011 in die UNESCO-Liste des Weltkulturerbes aufgenommene Museo di Santa Giulia, beheimatet einen Parcours, der durch Geschichte, Kunst und Spiritualität der Stadt und der Region führt. Angefangen von der Frühgeschichte, bis hin zum 19. Jahrhundert. Unter anderem findet sich auch typisch italienische, klassische Kunst sowie Nachkriegskunst.

Im Zeichen der Kooperation mit der Daimler AG konnte das Museum vom 7. März bis zum 30. Juni 2013 einen weiteren Kulturschatz ausstellen. Unter dem Titel „Novecento Mai Visto. Highlights from the Daimler Art Collection from Albers to Warhol to (now)" präsentierte die Daimler AG einen Teil ihrer umfangreichen Kunstsammlung, mehr als 230 Werke, die zwischen 1909 und 2013 entstanden, und von rund 110 internationalen Künstlerinnen und Künstlern angefertigt wurden. Darüber hinaus zeigte die Ausstellung auch zeitgenössische Installationen, Objekte, Fotografie und Videokunst, unter anderem von John M. Armleder und Martin Boyce sowie die berühmte Serie „Cars" von Andy Warhol. Parallel dazu hatte das Museo Mille Miglia drei Fahrzeuge aus der Sammlung von Mercedes-Benz Classic mit Bildern von Andy Warhol künstlerisch in Szene gesetzt. Im Innenhof des Museums wurden ein W 125, ein 300 SL „Gullwing" und C 111 in modernen Schaukästen präsentiert. •

Mercedes-Benz and the northern Italian region of Brescia, home to the Mille Miglia, are united by a shared passion for historic motor sport and art. While the race now belongs to history, Daimler hasn't forgotten Mercedes-Benz's role in the Mille Miglia legend and is launching a new partnership with two museums in Brescia, Museo di Santa Giulia and Museo Mille Miglia. The collaboration aims to highlight Mercedes-Benz Classic's links with the region and with the Mille Miglia – probably europe's most famous and most historic open-road race.

Mercedes-Benz is the only non-Italian automaker to have won the thousand-mile race twice on the original route. Rudolf Caracciola and Wilhelm Sebastian were responsible for the first victory in 1931 in a Mercedes-Benz SSK. Then in 1955, Stirling Moss and Denis Jenkinson won the rally in a record time that remains unbroken. The new collaboration is all about celebrating racing history and the region's heritage.

In 2011, UNESCO added Carl Benz's patent for the automobile to its "Memory of the World" register and also named the Santa Giulia monastery a World Heritage Site. The Museo di Santa Giulia, located inside the monastery, guides visitors through the history, art and spirituality of the town and surrounding region, from thousands of years ago through to the 19th century. The museum's collection also includes classical Italian art and post-WWII art.

As part of its collaboration with Daimler, the Museo di Santa Giulia is holding a special exhibition from March 7 to June 30, 2013. Entitled "Novecento Mai Visto. Highlights from the Daimler Art Collection from Albers to Warhol to (now)", the exhibition features over 230 works by 110 artists from around the world, spanning a timescale from 1909 to the present day. The exhibition also includes contemporary installations, „objets d'art", photography and video art, including pieces by John M. Armleder and Martin Boyce, as well as Andy Warhol's famous "Cars" series.

In parallel to this, Museo Mille Miglia has borrowed three automobiles from the Mercedes-Benz Classic collection and is exhibiting them against special backdrops of pictures by Andy Warhol. A W 125, a 300 SL Gullwing and a C 111 are currently being showcased in display boxes in the museum's inner courtyard. •

Max Ackermann
Chromatisch räumlich, 1937
Öl auf Hartfaserplatte
167 x 76 cm
Max Ackermann
Chromatisch räumlich
(Chromatic Spatial), 1937
Oil on hardboard
167 x 76 cm

Willi Baumeister
Ruhe und Bewegung II (auf Blau), 1948, Öl
mit Kunstharz auf Hartfaserplatte
81 x 100 cm
Willi Baumeister
Ruhe und Bewegung II (auf Blau)/Calmness
and Motion II (on Blue),
1948, oil with synthetic resin on hardboard
81 x 100 cm

Ulrike Rosenbach
Art is a Criminal Action, 1972/96
Fotomontage, digital bearbeitet (Diasec), Ed. 4/6
171 x 150,5 cm
Ulrike Rosenbach
Art is a Criminal Action, 1972/96
Photomontage, digitally processed
(Diasec), Ed. 4/6
171 x 150.5 cm

Maria Lemmé
Frauen in einer Landschaft, 1930
Pastell auf Leinwand
55x47x4 cm
Maria Lemmé
Women In A Landscape, 1930
Pastel on canvas
55 x 47 x 4 cm

Robert Longo
Untitled (engine), 1995
Kohle auf Papier auf Masonite/
charcoal, graphite on paper
mounted on masonite
152,5 x 152,5 x 0,5 cm
Robert Longo
Untitled (engine), 1995
Graphite on paper mounted on
masonite
152.5 x 152.5 x 0.5 cm

Richard Paul Lohse
Eine und vier gleiche Gruppen,
1949/1968
Öl auf Leinwand
120 x 120 x 3 cm
Richard Paul Lohse
Eine und vier gleiche Gruppen
(One and four identical groups),
1949/1968
Oil on canvas
120 x 120 x 3 cm

[redacted]

Jane Alexander
African Adventure, 2000, Digitaler Tintenausdruck auf Baumwollpapier, 235 x 340 cm

Jane Alexander
African Adventure, 2000, Digital ink print on cotton paper, 235 x 340 cm

BOM

WORLD
Diebels Alt
AMG
BAUMLER
men's fashion
BOSCH
Shell
SONAX

MIT DEM EVO II DURCH DIE GRÜNE HÖLLE

TRAVERSING "THE GREEN HELL" IN THE EVO II

NÜRBURGRING, NORDSCHLEIFE. MIT ROLAND ASCH AUF DEN SPUREN EINER ERFOLGSGESCHICHTE.

THE NORTH LOOP OF THE NÜRBURGRING. ROLAND ASCH RETRACES HIS SUCCESS.

Roland Asch ist beim Rennfahren mit dem schwarzen EVO II immer noch einer der ganz Großen.

Roland Asch is still one of the very best when racing in his black Evo II.

Nürburgring. Nordschleife. „Grüne Hölle". Draußen ist es, wie man hier so sagt, „saukalt". Die Wolken über dem historischen Boden hängen tief, wirken stellenweise schon fast erdrückend und brechen erst im Laufe des Rennens etwas auf. Ein leichter Nieselregen benetzt den Fahrbahnbelag, lässt dort leichten Dampf aufsteigen, wo sich Sekunden vorher noch Reifen am rauen Asphalt gerieben haben. Ein Mercedes-Benz 190 E 2.5-16 Evolution II (W 201) prescht den Streckenabschnitt „Hohe Acht" hinauf und auf der anderen Seite noch schneller wieder hinunter. Trotz Nässe scheint er fest am Boden zu kleben. Am Steuer Ex-Mercedes-Benz DTM Fahrer Roland Asch, der den Youngtimer-Boliden mit einem Affenzahn über die berühmte Rennstrecke jagt. Er ist ganz in seinem Element, denn ihn verbindet mit dem Evo II, wie Fans den Mercedes-Benz 190 E 2.5-16 Evolution II kurz und liebevoll nennen, eine ganz besondere Geschichte.

Heute, bei der Westfalen Trophy wird sie vor seinen Augen wieder lebendig und es fühlt sich an wie zu seinen ganz großen Zeiten, das war Anfang der 1990er. Roland Asch ist beim Fahren immer noch einer der ganz Großen. Auf der Strecke wird er eins mit seinem Evo II. Immer, wenn er an einer Menschenmenge vorbeifährt, schauen ihm die Leute nach. Wie früher. *„Der Evo II war damals ein Auto, das man sich von Mercedes-Benz eigentlich gar nicht vorstellen konnte. Flügel, Heckspoiler, das kannte man von anderen Sportwagen, aber doch nicht von Mercedes!"*, erinnert sich der ehemalige Mercedes-Benz Werksfahrer und DTM-Pilot. *„Den fuhr ich sogar privat. Der war so außergewöhnlich, dass ich einige Male von der Polizei angehalten wurde."* Die dachte damals, der Evo II sei aufgemotzt, kein normales Auto. *„Sie konnten einfach nicht glauben, dass der imposante Heckspoiler Serie war. Ich sagte denen, den kann man so ab Werk kaufen, das ist Original! Nachdem ich sie mit den Unterlagen überzeugt hatte, ließen sie mich weiterfahren."* Laut Asch meist mit einem Lächeln im Gesicht, denn die Polizisten fanden den kleinen supersportlichen „Baby-Benz" meist ziemlich toll. Und viele von ihnen erkannten auch Roland Asch sofort: *„Sie fahren doch bei der DTM, oder?"* Heute ist Roland Asch nicht mit seinem privaten Evo II unterwegs: *„Der steht sicher verstaut zu Hause in der Garage. Er ist Baujahr 1990 und hat nur 120.000 Kilometer runter. Alles ist noch wie früher, alles original! Sogar der erste Auspuff ist noch dran. Da klappert nichts, da rappelt nichts! Ein Traum. Und ab und an fahre ich ihn auch auf dem Nürburgring."*

Werfen wir aber wieder einen Blick auf den Track, die Performance von heute. Der ehemalige DTM-Fahrer muss schnell runterschalten, denn er steuert rasant die nächste Kurve an, der Motor des Rennboliden spielt auf, und er lenkt den Rennwagen aus der Kurve, raus auf die nächste Gerade. All das wirkt spielend leicht, ist es aber bei Weitem nicht. *„Die komplexe Mechanik, die ausgefeilte Lenkung, dass Doppelkuppeln, Zwischengas geben, das ist Rennfahren, das ist der Wahnsinn!"* Asch liebt das. Noch immer. Er hat großen Respekt vor den Leistungen der Mitstreiter und Konkurrenten aus der vergangenen Zeit. *„Damals sind wir ja noch richtig gegeneinander angetreten, sind noch lange Rennen gefahren, nicht nur halbstündliche, wie man es heute oft hat, das ging oft stundenlang. Das verdient Respekt"*.

Bewundern darf man auch den Elan und die Sportlichkeit, die Roland Asch mit seinen über 60 Jahren noch an den Tag legt.

The scene: the North Loop of the Nürburgring, or "The Green Hell", as the historic racetrack was nicknamed by Jackie Stewart. It's a cold day, and an oppressive sheet of low cloud blocks out the sky, only breaking up a little once the race is well underway. Light drizzle soaks the asphalt, quickly evaporating to steam in the still-hot areas the cars have just raced over. A Mercedes-Benz 190 E 2.5-16 Evolution II (W 201) accelerates into the steep "Hohe Acht" bend, coming out at breakneck speed on the downhill side yet managing to keep a firm grip of the track despite the wetness. Behind the wheel of this modern classic: former Mercedes-Benz DTM driver Roland Asch. He's completely in his element – he and the Evo II, as fans call the Mercedes-Benz 190 E 2.5-16 Evolution II, have a long history together.

Today, the day of the Westfalen Trophy, Asch's eyes are full of life and he feels just the way he did back in the early 1990s, when he was a major motorsport star. He's still a phenomenal driver – on the racetrack, he becomes one with his Evo II, and gazes follow him whenever he speeds past, just as they did then. *"At the time, the Evo II was a car you couldn't imagine was a Mercedes-Benz. The wings and rear spoiler looked like the type you might find on other sports cars, but not on a Mercedes!"*, recalls the former Mercedes-Benz factory driver and DTM racing driver. *"I even drove the Evo II as my private car. It was so unusual that the police stopped me on quite a few occasions."* They thought the Evo II was illegally modified. *"The rear spoiler was so striking they couldn't believe it was part of the standard build"*, Asch explains. *"I told them it was original, that you could buy the car like that straight from the factory. Once I showed them the documentation to prove this, they let me go."* Normally with a smile on their faces – Asch says the police seemed to love his compact, sporty "Baby Benz". Many of them also recognised Asch immediately and asked him *"You race in the DTM, don't you?"*. Today, he no longer drives his Evo II on a regular basis. *"It's stored securely in my garage at home"*, says Asch. *"It was made in 1990 and only has 120,000 kilometres on the clock. It's 100% original, everything is just the way it always was – it even has the original exhaust with no rattle or vibration. It's a dream. I still race it on the Nürburgring from time to time."*

Back to the track. The former DTM driver is heading for the next bend at high speed and needs to quickly shift down a gear. The Evo II's engine revs, and Asch steers the car out of the bend onto a straight section of track. He makes it look so easy, but it's anything but. *"Complex mechanics, nuanced steering, dual-clutch transmission, double declutching – that's what racing's all about. It's crazy!"* This is what Asch loves, and he thinks highly of his racing colleagues and competitors from the 1990s. *"In those days, we competed against each other in long races – not just half-hour ones like today, but races lasting for hours. I respect anyone able to race for that long."*

Now in his 60s, Asch is still dazzlingly sporty and energetic. During the day, he works in his garage, and the rest of the time he races as often as possible. *"I want to continue racing for as long as I can"*, he says. His Evo II was specially rebuilt by Mercedes-Benz Classic for retro car racing, and is fitted with a series-produced engine and an optimised undercarriage and braking system. *"I reckon there were around 150 to 200 cars in today's race"*, guesses Asch, happy to have been up against so much enthusiastic competition. *"Including a lot of big racing cars."* Not a problem for a racing

Tagsüber arbeitet er in seinem Autohaus, in der restlichen Zeit, die er zur Verfügung hat, lässt er den Rennfahrer aus Passion aufleben, und das will er auch weiterhin: *„Solange es eben noch geht."* Der Evo II, in dem er jetzt über die Nordschleife rast, ist von Mercedes-Benz Classic eigens für Youngtimer-Rennen frisch aufgebaut worden. Mit einem Serienmotor, optimiertem Fahrwerk und Bremssystem. *„Zwischen 150 und 200 Autos schätze ich, sind heut an Start gegangen"*, freut sich Asch über die Konkurrenz im Feld. *„Auch viele große Rennwagen."* Aber für den „alten Haudegen" alles kein Problem, er sieht das als sportliche Herausforderung. – Runter und vorbei am Sektor „Breitscheid", geht es Richtung „Kesselchen". Asch schaltet höher und gibt Gas. Der kräftige Motor des Evo II dreht sonor auf, spielt sein kerniges Beschleunigungslied. Dynamisch. Forte. Mann und Maschine lassen die Konkurrenz locker hinter sich.

„Wenn ich meine aktive Rennfahrerzeit Revue passieren lasse und dann an den Evo II, zurückdenke, dann weiß ich, das war mein Auto." Es ist ein bisschen so wie bei der ganz großen Liebe. Gebaut wurden von ihm nur 500 Stück. Und der Evo II war selbst für damalige Verhältnisse nicht günstig. 115.000 D-Mark, das war 1990 der Preis. Dafür hielt sich der Verbrauch erstaunlich in Grenzen. Asch, ein Meister seines Fachs, schafft es zum Nürburgring im privaten Evo II mit nur einer Tankfüllung hoch und runter, das sind insgesamt etwa 680 Kilometer. *„Das glauben Sie nicht, ich fahre auf 100 Kilometer unter zehn Liter mit diesem Auto."* Der Rennfahrer hat sogar bei der Entwicklung mitgewirkt: *„Das war damals bloß eine kleine Truppe bei Mercedes. Drei Ingenieure, Läpple, Bulling, Seibold und ich."* Damals war Roland Asch Testfahrer auf einem Mercedes-Benz 190 und auch in seiner Freizeit hörte sein Interesse nicht an der Entwicklung des Evo II auf. *„Wir haben zum Beispiel das Aktivfahrwerk entwickelt und sogar das erste Renn-ABS."* Das konnte Asch gleich 1990 ausprobieren, denn die DTM hatte einen Gastauftritt in Südafrika. *„Das war das erste Rennen mit ABS und ich hab es gleich gewonnen, das war für mich ein besonderes Erlebnis."* Denn er hatte bei der Entwicklung maßgeblich mitgeholfen. *„Für mich war das alles kein Problem, denn ich bin in einer Werkstatt aufgewachsen, war schon immer praktisch veranlagt."* Und das muss man als Rennfahrer auch sein, denn schnelle Reaktionen, gute Intuition und Improvisationstalent sind Grundvoraussetzungen für den Erfolg.

Riesenerfolg hatte Roland Asch 1993 mit der letzten Ausbaustufe des Evo II. „Doppelsieg in Berlin!" Dann kam die Ansage, dass der Evo II ins Museum kommt. Asch war geschockt, konnte das erst nicht glauben. *„Als ich das gehört habe, das war schon brutal für mich."* Dann kam der Nachfolger, die C-Klasse und wurde zum erfolgreichsten DTM Fahrzeug überhaupt. *„Umso mehr freue ich mich, dass man das heute alles wieder aufleben lässt und das Auto wieder auf die Rennstrecke bringt."*

Asch lässt es bei dieser Westfalen Trophy auf dem Nürburgring selbst wieder richtig

veteran – he loves the challenge. Asch races downhill past the Breitscheid section of the North Loop towards the stretch known as the "cauldron". He shifts up a gear and accelerates. The Evo II's powerful motor roars sonorously, singing its song of acceleration loudly and with gusto. Man and machine leave the rest of the field behind them with ease.

"When I look back to my time as a racing driver and think of the Evo II, I'm positive it was the right car for me", says Asch fondly. It's like falling in love – you just know. Only 500 units of the Evo II were made, and they weren't cheap for the time: the car's price tag in 1990 was 115,000 Deutsche marks. As if to help recoup the purchase cost, the car's fuel consumption is remarkably low. Asch, who knows the car inside out, was able to drive his Evo II to the Nürburgring and back – a 680-kilometre journey – on a single tank of fuel. *"You won't believe me, but I can drive 100 kilometres on less than ten litres of fuel in this car"*, he explains.

Asch also played a major role in the car's development. *"There was a small group of us at Mercedes back then – three engineers, Läpple, Bulling and Seibold, and me."* As well as test-driving the Mercedes-Benz 190, he was passionate about it in his leisure time too. *"We developed the active undercarriage, for example, and even the first ABS in a racing car."* Asch was able to test the ABS out in 1990 when the DTM held an event in South Africa. *"It was the first race with ABS, and I won – that was a brilliant experience. Working on the car's development wasn't difficult, because I grew up in a car garage and have always been a practical person."* As a racing driver, you have to be – rapid reaction, excellent intuition and the ability to improvise are key to success.

Asch brought home more trophies in 1993, in the final, most powerful version of the Evo II. "A double win in Berlin!" he recalls proudly. Then came the announcement that the Evo II would be retired. A shocked Asch couldn't believe this at first. *"When I heard the news, I was devastated."* But the Evo II's successor, the C-Class, did its legendary forerunner proud by becoming the most successful DTM vehicle ever. Asch, though, has stayed loyal to the Evo II. *"I love that today, people still enjoy driving and even racing older cars like the Evo II"*, he remarks. And while he and the Evo II are certainly impressive when going for broke on the Nürburgring, they also attract a lot of attention off the track, including before and after the race and in the paddock. *"People always gather when I arrive somewhere with the car. And someone always asks 'Can I sit in it?' The car's so popular, the attention could go on all day."*

Asch steps on the gas again as he races over the "Döttinger Höhe", a long, straight, slightly elevated section of the track. The acceleration, the traction, and the stark sensation of the road beneath you – the driver feels almost every bump – are real adrenaline-boosters. *"We all started at the same time, so during the race you end up in a convoy every three minutes. And the whole time, you're driven by the desire to be in pole position, to win. Because when you do something, you want to do it properly – you want to win."* It's this thirst for victory that took Asch and his 190 E 2.5-16 Evolution II to

Den AMG-Mercedes 190 E 2.5-16 Evolution II entwickelten die Motorsportingenieure für die DTM Meisterschaften der Jahre 1990 bis 1992 auf Basis der 190 E 2.3-16. Mit dem „Evo II“ holte Klaus Ludwig 1992 den ersten Meisterschaftstitel der DTM für Mercedes-Benz.

Mercedes-Benz‘s motorsport engineers developed the AMG-Mercedes 190 E 2.5-16 Evolution II for the 1990-1992 DTM race series based on the 90 E 2.3-16. Klaus Ludwig won the DTM for Mercedes-Benz for the first time in 1992 in the Evo II.

[redacted]

krachen. Aber auch abseits der Piste sorgen er und „sein“ Evo II für eine Menge staunender Gesichter. Egal ob am Vorstart oder rund um das Rennen, etwa, wenn er mit seinem Evo II am Fahrerlager hält, ist immer was los. *„Im Prinzip gibt es überall Menschentrauben, wenn ich mit dem Auto ankomme. Und ich werde immer wieder angesprochen: ‚Darf ich mal reinsitzen?' Das kann den ganzen Tag so gehen, so beliebt ist das Auto.“*

An der Döttinger Höhe vorbei gibt der ehemalige DTM-Fahrer noch mal richtig Gas, will es wissen. Bei der Beschleunigung, der Traktion und dem puristischen Gefühl unter dem Hintern – man spürt fast jede Bodenwelle – treibt es einem den Adrenalinspiegel nach oben. *„Wir sind alle miteinander gestartet. Alle drei Minuten ein Pulk. Begleitet vom Gefühl, dass du einfach immer vorne sein, gewinnen willst. Denn wenn du etwas machst, willst du es richtig machen, willst gewinnen.“* Und der Siegeswillen ist es, mit dem der Rennfahrer seinen 190 E 2.5-16 Evolution II an die Spitze treibt. *„Mir macht das richtig Spaß, vor den Gegnern zu fahren. Technisch gesehen sind ja viele eigentlich schneller und stärker.“* Aber mit dem Evo II, dem fahrerischen Können von Roland Asch und der richtigen Bereifung liegen gleich drei Trümpfe auf der Hand. *„Es ist zwar feucht, aber das Auto lässt sich unheimlich gut fahren, auch wenn die anderen mit Regenreifen unterwegs sind, habe ich mich bewusst für Slicks entschieden. Das war eigentlich das Beste für mich und das Auto. Für andere war es wohl eher Risiko.“*

Eine weitere Eigenart von Asch: er denkt beim Rennfahren nie nur an sich, sondern immer auch an sein Auto. Er schätzt den Evo II, er ist sein ebenbürtiger Partner und so behandelt er ihn auch. *„Wenn es den Berg runtergeht, Richtung Breitscheit oder Hatzenbach, dann spüre ich, dass ich das Auto super im Griff habe. Ich muss mich wohlfühlen, dann passt alles.“* Für einen Rennfahrer sei es immens wichtig, dass er das Auto im Griff habe und nicht das Auto ihn.

Schnell und zackig schaltet Asch auf dem nächsten geraden Stück Asphalt wieder einen Gang hoch und treibt Drehzahlmesser und Tachonadel nach oben. *„Manchmal gibt es Rennwagen, die sind total überfüllt mit Technik und du bekommst nicht mehr die Rückmeldung, die du dir erhoffst. Dann weißt du nicht, woran du bist. Ich kann dann nicht schnell fahren.“* Wie wichtig das ist, konnte Asch beim Abschnitt „Flugplatz “ zeigen, nun ist Feingespür und Können gefragt, um auch bei hohen Geschwindigkeiten nicht abzuheben. *„Da musst Du die Fahrphysik spüren und kennen: Da bist Du schnell, da kommst Du mit dem Vierten oder Fünften über die Kuppe, dann rechts, da musst Du das Auto voll im Griff haben.“* Und dass dies bei der diesjährigen Westfalen Trophy der Fall war, konnte Roland Asch nach einem harten Rennen durch seine Gesamtplatzierung untermauern: Rang sieben unter den verschiedenen Klassen mit teilweise stärkeren Motoren, das kann sich sehen lassen.

Als Roland Asch aussteigt, beantwortet er noch schnell die Fragen eines Fans: *„Ich schätze so 240 fährt er, vielleicht etwas schneller.“* Genaue km/h-Zahlen waren nie wirklich wichtig. Erfolge um so mehr. Und die konnte der Evo II zuhauf vorweisen, auf den Rennstrecken der DTM und bis heute in den Herzen

the pinnacle of success. *"I really enjoy being ahead of the others, because many of them are in cars that are more powerful and can reach higher speeds than mine."* Another factor in Asch's strong performance in the Westfalen Trophy was his choice of tyres. *"Although it's wet, I have excellent control. The others are driving on rain tyres but I opted for slicks – that was the best choice for me and the car. For others, it might have been too big a risk."*

Asch also stands out from the field due to his relationship with the Evo II, which he values as an equal partner, not just a tool. *"On downhill stretches, like the ones before the Breitscheid and Hatzenbach sections of the North Loop, I always notice what great control of the car I have. I have to feel comfortable in the car to perform at my best."* For racing drivers, it's important that they control the car, not vice versa.

Reaching the next straight section of track, Asch quickly shifts up a gear again, and the speedometer and tachometer climb. *"You sometimes get racing cars that are so full of technology, you don't get the feedback you want from them. Then you don't know what you're doing. I can't drive at high speeds in a car like that."* Asch demonstrates how important this is in the section of the North Loop nicknamed the "airfield" – a stretch of track that demands skill and sensitivity behind the wheel due to the risk of losing grip at high speed. *"You've got to feel the physics, the dynamics, and know what you're doing – know when to accelerate, know when to shift to fourth or fifth gear to climb over the crest before turning right... you've got to have complete control of the car."* Evidently, Asch did in this year's Westfalen Trophy, finishing a very strong seventh overall after a tough race. An especially impressive placing considering many of the other cars had more powerful engines.

When Asch gets out of his car at the end of the race, one of the first things he does answer a fan's question about his speed. *"I reckon about 240 km/h, maybe a bit more."* Knowing his exact speed has never been important to Asch; it's doing well he cares about. And just as the Evo II brought him triumph after triumph two decades ago in the DTM, it catapults him into the affections of retro car lovers here at the Nürburgring. Famously available with either leather or tartan seats, the Evo II was celebrated in the late 1980s and throughout the 1990s, but faded from the public consciousness after the millennium. Now, it's experiencing a renaissance in popularity. More and more people are showing an interest in retro cars, and as a result, Mercedes-Benz Classic is stepping up its classic car racing activities. As well as Asch's Evo II, Mercedes-Benz Classic was also represented in this year's Westfalen Trophy by a Mercedes-Benz 220 SE (W 111) – the Racing Fintail. Rebuilt in 2011, the car regularly competes in races organised by the German Drivers' Association for Historic Racing.

At the end of the long race day, Roland Asch has one more wish. *"It'd be great if sometime soon, there were a racing series just for the Evo II."* Time will tell whether this happens. For now, Asch has already entered next year's Westfalen Trophy in his calendar. When the date arrives, though, he might not be competing in the Evo II – because Mercedes-Benz Classic is busy restoring other classic models to be raced.

Beim Rennen der Fahrergemeinschaft historischer Rennsport e.V. (FHR) in Spa-Francorchamps (September 2012) erzielte der W 111 mit Roland Asch am Steuer einen Klassensieg.
In the September 2012 race at Spa-Francorchamps organised by the German Drivers' Association for Historic Racing, Roland Asch drove the W 111 to victory.

„Grüne Hölle“ wird die Nordschleife des Nürburgrings genannt, auf der jedes Jahr das Westfalen-Rennen der Youngtimer Trophy stattfindet.
„The Green Hell“ is the nickname Jackie Stewart gave to the North Loop of the Nürburgring, which hosts the annual Westfalen Trophy for retro cars.

Genaue km/h Zahlen waren Roland Asch nie wichtig, Erfolge umso mehr. Und die gab es mehrfach.
Knowing his exact speed was never important to Asch; it's winning he cares about. And winning is what he's good at.

[redacted]

vieler Youngtimer-Fans. Ob mit Leder oder den berühmten Sitzen im Schottenkaro-Stoff. Der Evo II erfreute sich Ende der 1980er und in den 1990er Jahren großer Beliebtheit, geriet dann etwas in Vergessenheit und erlebt nun seinen zweiten schnellen Frühling. Man hat längst die Zeichen der Zeit erkannt, die Youngtimer-Fangemeinde wächst und mit ihr das Engagement von Mercedes-Benz Classic im historischen Motorsport.
Neben dem Evo II, in dem Roland Asch unterwegs war, schickte Mercedes-Benz Classic auf der diesjährigen Westfalen Trophy auch einen Mercedes-Benz 220 SE (W 111), die sogenannte „Rennflosse", auf die Piste. Das 2011 neu aufgebaute Fahrzeug ist regelmäßiger Vertreter im Rennzirkus um die FHR – die Fahrergemeinschaft historischer Rennsport e.V.

Am Ende dieses langen Renntages hat Roland Asch noch einen ganz besonderen Wunsch auf dem Herzen. *„Es wäre schön, wenn für den Evo II bald eine eigene Rennserie existieren würde."*

Ob das passiert, wird sich zeigen. Fest steht, auch im kommenden Jahr will er wieder die „Grüne Hölle" bezwingen. Ob im Evo II ist noch unklar, denn Mercedes-Benz Classic arbeitet daran, weitere Klassiker fit zu machen und wieder ins Rennen zu schicken. •

Der Mercedes-Benz Typ 220 SE (W 111) wurde von Mercedes-Benz Classic für den historischen Rennsport aufgebaut.
The Mercedes-Benz 220 SE (W 111) was specially rebuilt by Mercedes-Benz Classic for retro car racing.

[redacted]

„Den EVO II fahre ich sogar privat. Der ist so außergewöhnlich, dass ich einige Male von der Polizei angehalten wurde.“ Roland Asch
“I even drove the Evo II as my private car. It was so unusual that the police stopped me on quite a few occasions“, recalls Asch.

DREIFACHSIEG MIT DREIFACHSIEGER
A TRIPLE VICTORY AND TRIPLE VICTORS

SILBERPFEILE TRIUMPHIERTEN AUF DEM HOCHGESCHWINDIGKEITSKURS VON TRIPOLIS.

SILVER ARROWS TRIUMPHING AT THE HIGH-SPEED TRACK IN TRIPOLI.

Startvorbereitungen am Mercedes-Benz W 165 beim traditionsreichen Rennen in Tripolis 1939.
Preparing the Mercedes-Benz W 165 at the classic race in Tripoli in 1939

[redacted]

Auch heute noch bemerkenswert: Seltene Farbaufnahmen der 1930er-Jahre zeigen die legendären Silberpfeile beim Großen Preis von Tripolis.
Still striking today: rare colour photos from the 1930 show the legendary Silver Arrows at the Grand Prix in Tripoli.

Ungeachtet der erfolgreichen Vorgeschichte stand der 7. Mai 1939 für die Fahrer von Mercedes-Benz eigentlich unter keinem guten Stern. Ein Jahr zuvor hatten Hermann Lang, Manfred von Brauchitsch und der spätere Europameister, Rudolf Caracciola, auf dem Hochgeschwindigkeitskurs von Tripolis triumphiert und mit ihren Silberpfeilen vom Typ W 154 einen ungefährdeten Dreifachsieg eingefahren. In den mittleren und späten 1930er Jahren galt das nahe der libyschen Hauptstadt gelegene Autodromo della Mellaha als schnellster Kurs der Welt. Und auch, wenn der Grand Prix nicht zu den offiziellen Wertungsläufen des damaligen Formel-1-Pendants gehörte, wollte jeder Hersteller das prestigeträchtige Rennen in der italienischen Orientkolonie gewinnen. Ohnehin dominierte Mercedes-Benz das Renngeschehen in Europa nahezu nach Belieben. Für die motorsportbegeisterten Italiener eine Schmach. Deshalb änderte Mussolinis Motorsportbehörde das Reglement für alle Rennen auf italienischem Boden vermeintlich zugunsten von Alfa Romeo und Maserati: Der Hubraum wurde auf 1.500 Kubikzentimeter, der sogenannten „Voiturette-Formel“, entsprechend begrenzt.

Für Mercedes-Benz und seinen Rennleiter Alfred Neubauer bedeutete das, dass sie rasch reagieren und einen wettbewerbsfähigen Silberpfeil auf die Räder stellen mussten. Wie rasch die Entwicklung des neuen, nur halb so großen Aggregats vonstattengehen musste, zeigt ein Blick in die Annalen der Motorsportabteilung von Mercedes-Benz: Am 11. September, unmittelbar nach dem Gran Premio von Monza, erfuhren die Mannen um Neubauer von den einschneidenden Änderungen. Bis zum Rennen von Tripolis, dem Saisonstart des Jahres 1939, blieben knapp acht stressige Monate. Daher musste der neue W 165 genannte Bolide auf dem Vorgängerfahrzeug aufbauen. Der kleine, nur 195 Kilogramm leichte V8 schöpfte 254 PS (187 kW) aus nur 1.495 Kubikzentimetern Hubraum. 1938 leisteten die von Chefingenieur Rudolf Uhlenhaut entwickelten V 12 Silberpfeile noch 474 kompressoraufgeladene PS (349 kW). Doch die Italiener, die seit 1934 danach trachteten, die Vormachtstellung von Mercedes-Benz und Auto Union auf dem Kurs von Tripolis zu brechen, gingen erneut leer aus: Wieder gewann Mercedes-Benz in der arabischen Gluthölle.
Allen voran fuhr erneut Hermann Lang auf das Treppchen, gefolgt von Caracciola. Für Lang ist es bereits der dritte Erfolg in Tripolis. Der „Lieblingsschüler“ von Motorsportchef Neubauer begann seine Karriere bei Mercedes-Benz als Mechaniker in der Rennabteilung. Da er „Benzin im Blut“ hatte und über reichlich Talent verfügte, gehörte er bald zur erfolgreichen Mannschaft um Manfred von Brauchitsch und Rudolf Caracciola. Die Strecke in Italienisch-Libyen lag dem Schwaben, weil sie anspruchsvoll war und höchstes Tempo ermöglichte. So „trug sich“ Lang drei Mal hintereinander – 1937, 1938 und 1939 – als Gewinner in die Annalen ein. Im letzten

Despite Mercedes-Benz drivers' previous history of successes, the stars were not particularly well-aligned for them on 7 May 1939. The year before, Hermann Lang, Manfred von Brauchitsch and Rudolf Caracciola (who later went on to become European Champion) had triumphed on the race track at Tripoli, and they had led their W 154 Silver Arrows to an uncontested triple victory. In the mid- to late 1930's, the Autodromo della Mellaha near the Libyan capital was seen as the world‘s fastest racetrack. And even though the Grand Prix was not officially part of what was then the equivalent to Formula One, every automotive manufacturer still wanted to win the prestigious race in the Italian colony. In the world of European motorsport, Mercedes-Benz won practically whenever it chose to, which Italian motorsport enthusiasts saw as an outright scandal. Consequently, Mussolini‘s motorsport authorities modified the rules for all races that took place on Italian soil, presumably to give an advantage to Alfa Romeo and Maserati: the engine size was limited to 1,500 cc's, in keeping with the voiturette formula.

When the rules changed, Mercedes-Benz and its racing manager Alfred Neubauer had to react quickly and put together a Silver Arrow that could hold its own. A glance at the annals of Mercedes-Benz's motorsport department shows how quickly they had to develop a new car that was only half the original size. On 11 September, immediately after the Gran Premio of Monza, Neubauer‘s team learned about the game-changing new guidelines, which left them less than eight stressful months until the race in Tripoli that would kick off the 1939 season. As a result, the new model, which was named W 165, had to be based on its predecessor. At only 195 kilogrammes, the light V8 drew 250 hp (187 kW) despite an engine capacity of a mere 1,495 cubic centimetres. By 1938, the V12 Silver Arrow designed by chief engineer Rudolf Uhlenhaut featured 467 supercharged hp (349 kW). But the Italians, who had been attempting to end Mercedes-Benz and Auto Union‘s winning streak on the Tripoli track since 1934, went home empty-handed that year: Mercedes-Benz was the victor in the sweltering Arab venue once again.
And as before, Hermann Lang took the top spot on the winner‘s rostrum, followed by Caracciola. It was already Lang‘s third victory in Tripoli. As the “teacher's pet” of motorsport head Neubauer, Lang started his career as a mechanic in the race-car department at Mercedes-Benz. Since he was immensely talented and had “petrol in his blood”, he quickly made his way into the successful team with Manfred von Brauchitsch and Rudolf Caracciola. The track in Italian Libya suited Lang, since it was demanding and made it possible to drive at top speeds. He made history by going home with the trophy three years in a row – 1937, ‘38 and ‘39 – although in the final year his win took place under challenging circumstances. In addition to the new technical stipulations which were intended to thwart the Stuttgart team, there was also controversy surrounding the

Die Mehrzahl der Rennplakate von 1938 und 1939 wurden von Walter Gotschke angefertigt, der damit zu einem der bekanntesten Illustratoren rund um das Automobil wurde.

Most of the 1938 and 1939 racing posters were produced by Walter Gotschke, who became one of the automotive world's best-known illustrators.

Jahr freilich unter besonders erschwerten Bedingungen. Denn neben dem neuen technischen Reglement, das Siege der Stuttgarter verhindern sollte, war es die staatliche Lotterie, die für hitzige Gespräche sorgte: Für 12 Lire konnte man sowohl auf „Sieg“ als auch auf „Platz“ setzen – und bis zu drei Millionen Lire kassieren. Schon im Vorfeld wurde klar, dass außergewöhnlich viele Wetten für italienische Hersteller eingingen. Zu viele? Wurde manipuliert? Hartnäckig hielt sich das Gerücht, die Fahrer Achille Varzi, Tazio Nuvolari und Baconin Borzacchini seien in unlautere Machenschaften verwickelt.

In Stuttgart fragte man sich, ob man überhaupt gewinnen konnte. Und wie man konnte: Während Hermann Lang, wie vorher festgelegt, einen Boxenstopp einlegte, fuhr der amtierende Europameister Caracciola das Rennen durch. Lang siegte mit einer Runde Vorsprung und hätte den am Ende zweitplatzierten Markenkollegen sogar überrunden können. Dank längerer Übersetzung und enormem fahrerischen Geschick steuerte der Schwabe einem ungefährdeten Sieg entgegen. Den Dreifach-Triumph von 1938 konnte Mercedes allerdings nicht wiederholen, denn Dritter wurde Emillio Viloresi mit seinem Alfa Romeo 158 – der Italiener kam knapp acht Minuten nach Lang ins Ziel. Der pfeilschnelle W 165 brauchte für die 393 Kilometer oder 30 Runden von Tripolis 1 Stunde, 59 Minuten und 12 Sekunden. Beim Hattrick aus dem Vorjahr war Lang 2 Stunden, 33 Minuten und 17 Sekunden unterwegs gewesen – hatte allerdings 525 Kilometer bzw. 40 Runden zu

Italian state lottery: for 12 lire, you could bet both on “win” and on “place” to win up to three million lire. It became clear well in advance that an unusual number of bets would be placed on Italian manufacturers. Too many, perhaps? Was someone pulling strings? There were persistent rumours that drivers Achille Varzi, Tazio Nuvolari and Baconin Borzacchini were involved in shady business.

In Stuttgart, people were asking themselves if there was even any hope of winning, and if so, how. During the 1939 race, Hermann Lang took a pit stop as scheduled, but his Mercedes colleague Caracciola, the reigning European champion, drove without stopping. Nevertheless Lang outpaced second-place winner Caracciola and even ended up a full lap ahead. With the help of a longer gear ratio and his incredible talent behind the wheel, the German made his way to an unimpeded victory. The triumphant 1938 Mercedes-Benz trifecta was something the company was unable to repeat, however, since this time third place went to Emillio Viloresi with his Alfa Romeo 158: the Italian reached the finish line nearly eight minutes after Lang. The W 165 flew like an arrow, needing 1 hour, 59 minutes and 12 seconds for the 393 kilometres or 30 laps in Tripoli. By contrast, in 1938 Lang was en route for 2 hours, 33 minutes and 17 seconds, but he also had to cover 525 kilometres / 40 laps.

Strictly speaking, the Silver Arrows were not really a technical innovation. These winning cars, which were born of necessity, were actually enhanced models based on their success-

Gleich am Start des prestigeträchtigen Rennens übernahm Hermann Lang die Führung und gab sie bis zum triumphalen Doppelsieg nicht mehr her.
From the very outset of the prestigious race, Hermann Lang took the lead and did not let go until the triumphant double victory.

bewältigen. Ein echtes technisches Novum waren die Silberpfeile nicht. Die „aus der Not geborenen“ Sieger waren konsequente Weiterentwicklungen der erfolgreichen Vorgängermodelle. Jedes Jahr wurden neue Regelungen eingeführt, so entschied die Rennsportorganisation AIACR (Association Internationale des Automobile Clubs Reconnus) 1936, dass zur Saison 1938 Fahrzeuge mit Kompressor maximal drei Liter besitzen und ohne Aufladung 4,5 Liter nicht überschreiten durften. Alfred Neubauers Mannen waren also einiges gewohnt. Der dritte Sieg von Tripolis bot aber doch eine technische Innovation – allerdings an und neben der Strecke: Denn erstmals wurde „in Farbe“ fotografiert. Atmosphärisch dicht fangen etwa die Bilder mit Lang, Caracciola oder den jubelnden Zuschauern ein, wie man Motorsport 1939 „lebte“. Man taucht ein in die Historie, die vor den eigenen Augen lebendig wird, sieht die Crew beim Boxenstopp oder die Fahrer neben ihren Boliden. Und auch, wenn sich vieles verändert hat: Autorennen sind immer noch ein Spektakel, egal ob die Männer in den Boliden Lang, Caracciola, Fangio, Häkkinen oder Hamilton heißen. •

ful predecessors. New racing rules were established every year, and in 1936 the auto-racing organisation AIACR (Association Internationale des Automobile Clubs Reconnus) stated that in the 1938 season, cars were limited to engine displacement of 3 litres with a supercharger and 4.5 litres without. So Alfred Neubauer's team was used to handling quite a bit. Their third win in Tripoli involved a technical innovation, however, although this time it took place at and alongside the tracks: for the first time, colour photographs were taken. The pictures of people such as Lang, Caracciola or throngs of cheering fans clearly capture the rich atmosphere and the way people "lived" motorsport in 1939. Looking at these images means stepping back into history and watching it come to life before your eyes as you see the crew at the pit stop or the drivers next to their cars. And even though so much has changed, motorsport races are still as much of a grand spectacle as ever, no matter whether the drivers' names are Lang, Caracciola, Fangio, Häkkinen or Hamilton. •

LEGENDEN
DIE GESCHICHTE SCHRIEBEN
LEGENDS
WHO MADE HISTORY

4.

AGIP

SILBER
DIE FARBE DER SIEGER
SILVER
THE COLOUR OF VICTORY

RENNFAHRERLEGENDE JUAN MANUEL FANGIO UND DER „SILBERPFEIL."

RACING LEGEND JUAN MANUEL FANGIO AND THE SILVER ARROW.

Nach dem Rennen in Reims lassen sich Karl Kling (links) und Juan Manuel Fangio als Doppelsieger feiern.
Karl Kling (left) and Juan Manuel Fangio are celebrated after their victory in the Reims race.

[redacted]

Auf der AVUS belegt Juan Manuel Fangio im Mercedes-Benz W 196 R den 2. Platz hinter seinem Teamkollegen Kling.
In the AVUS race, Juan Manuel Fangio was 2nd in his Mercedes-Benz W 196 R behind teammate Kling.

Gleich der erste Einsatz der neuen Silberpfeile wird am 4. Juli 1954 zum Erfolg. Doppelsieg beim Großen Preis von Frankreich in Reims.
The new Silver Arrow triumphs in its very first outing: a double win in the French Grand Prix in Reims on July 4, 1954.

Die 1930er Jahre. Nirgendwo liegen Tragödien und Triumphe so dicht beieinander wie im Rennsport der damaligen Zeit. Die Szenerie Mann gegen Mann bei ungeahnt hohen Motor-Drehzahlen und Geschwindigkeiten ist die Begleitmusik zur Geschichte der legendären Silberpfeile von Mercedes-Benz. Sie gehören zu den neuen Helden eines heraufziehenden Technik-Zeitalters. Konstruktionen, die ihrer Zeit um Lichtjahre voraus sind.

1934 wird die neue „750-Kilo-Formel" eingeführt. Ein Rennwagen darf demnach beliebig stark sein, allerdings nur eine drei-viertel Tonne wiegen (ohne Treibstoff, Öl, Wasser und Reifen). Ziel ist die Drosselung der Höchstgeschwindigkeit. Die Antwort aus Stuttgart: der Mercedes-Benz W 25. Mit einem Achtzylinder-Motor, 3,4 Liter Hubraum, anfänglichen 354 PS (260 kW) und einer sensationellen Spitzengeschwindigkeit von fast 300 km/h bei offener Karosserie. So viel zum Thema Speed. Das Problem: Obwohl Mercedes-Benz Gewicht einspart, wo es nur geht, erreicht der W 25 nicht das vorgeschriebene Gewicht. Die Waage zeigt genau ein Kilogramm zu viel an! Zur Rettung wird eine simple Bemerkung des damaligen Fahrers Manfred von Brauchitsch: Seine Worte „Nun sind wir die Gelackmeierten!" bringen Rennleiter Alfred Neubauer der Legende nach auf die Idee, den weißen Lack abzuschleifen, um das Gewicht auf das zulässige Limit zu verringern. Über Nacht bringen die Mechaniker den Lack so weit herunter, bis das typisch silbern glänzende Aluminiumblech zum Vorschein kommt. Der silber glänzende Bolide wiegt nun exakt 750 kg, und prompt triumphiert Manfred von Brauchitsch mit seinem W 25 am 3. Juni 1934 beim Eifelrennen auf dem Nürburgring. Später teilt er der Presse mit: „Einen Silberpfeil zu fahren ist eine Ehre." Ein Mythos, der bis heute viele Menschen in den Bann zieht, ist geboren: die Mercedes-Benz Silberpfeile.

1937 setzt auch der neu entwickelte W 125 Maßstäbe, und nach der Einführung einer neuen Rennformel im Jahr 1938 setzt Mercedes-Benz seine Siegesserie mit dem 3-Liter-Rennwagen W 154 fort. In diesem Jahr wird der Begriff Silberpfeil zum ersten Mal als

The 1930s: Never before were tragedy and triumph more closely interlocked than in this decade's world of racing. The story of the legendary Mercedes-Benz Silver Arrow unfolded against a backdrop of man-to-man contests and the roar of unprecedented engine speeds and velocities. These were the new heroes of the burgeoning age of technology: models that were light-years ahead of their time.

In 1934 a new "750-kg formula" took effect. Racing cars were allowed to be as powerful as they liked, but they could only weigh up to three-quarters of a ton (excluding fuel, oil, water and tyres). The purported aim: to limit the maximum speed. Stuttgart's response: the Mercedes-Benz W 25. This open-body model featured an eight-cylinder motor, a cubic capacity of 3.4 litres, and 349 hp (260 kW) initially, reach-ing sensational top speeds of around 300 km/h—so much for the notion of limiting speed. The problem: although MercedesBenz shaved off mass wherever possible, the W 25 was unable to achieve the prescribed weight. The scales showed exactly one kilogram too many! A chance remark by driver Manfred von Brauchitsch saved the day: Legend has it that his glum words "Now we're painted into a corner!" gave racing director Alfred Neubauer the inspiration to meet the weight limit by removing the white paintwork from the car. Mechanics working through the night managed to strip off so much paint that the typical silver gleam of polished aluminium came shining through. The gleaming silver bolide now weighed exactly 750 kg. The result: Manfred von Brauchitsch drove the W 25 to a sensational victory on June 3, 1934 during the Eifel race at the Nürburgring. Later he would tell the press: "It is an honour to drive a silver arrow." Thus was born the legend that would captivate so many admirers to this day: the Mercedes-Benz Silver Arrow.

The W 125 model developed in 1937 set new standards, and when a new racing formula was issued in 1938, the three-litre W 154 continued the string of victories. That was the year when journalist Eberhard Hundt first used "Silver Arrow" as a collective term for all German racing cars. The cars that Auto Union

BOSCH
BOSCH

L'EQUIPE
LE CIMENT ARMÉ
LOP

Silberpfeile galten als automobile Konstruktionen, die ihrer Zeit um „Lichtjahre“ voraus waren.
Silver Arrows were seen as light years ahead of their time in terms of design.

Der Große Preis von Europa auf dem Nürburgring, 1954.
The European Grand Prix on the Nürburgring, 1954.

Sammelbegriff für die deutschen Rennwagen vom Journalisten Eberhard Hundt verwendet. Den Rennwagen der Auto Union bezeichnete man bereits zuvor aufgrund seiner aerodynamischen, silbernen Verkleidung als „Silberfisch". Die Auto Union, ein 1932 gegründeter Zusammenschluss der Autofirmen Audi, DKW, Horch und Wanderer, dominiert neben Mercedes-Benz bis 1939 das Grand-Prix-Geschehen, wobei sich Mercedes-Benz mit Rudolf Caracciola am Steuer 1935, 1937 und 1938 den Europameistertitel sichern kann – bevor man 1939 bei Kriegsausbruch nur innoffiziell den Titel holt. Immer wieder stellen die Spitzenfahrer der beiden Rennställe, wie Nuvolari, Stuck, von Brauchitsch, Caracciola, Lang oder Rosemeyer, in ihren Silberpfeilen neue Rekorde auf.

In Reims feiern die Stuttgarter im Juli 1954 den ersten Grand-Prix-Sieg seit 1939 – und das gleich beim ersten Rennen der neuen Silberpfeile. Dafür hatte man hart gearbeitet und einen komplett neuen Rennwagen nach der neuen 2,5-Liter-Formel entwickelt, der mit seiner futuristisch anmutenden Stromlinienkarosserie, mit dem Gitterrohrrahmen und der Benzindirekteinspritzung technische Maßstäbe setzt.
Zeitgleich ist das triumphale Comeback auch der Startschuss für einen Mann, dessen Name untrennbar mit Mercedes Benz verbunden sein wird: Juan Manuel Fangio.
Ein Mann, der außerhalb der „normalen" Formel-1-Fahrer-Riege steht. Cool, aber nicht kühl. Ruhig und überlegt, und vielleicht deshalb auch überlegen. Er dominiert die 1950er Jahre in der Formel 1. Er ist unglaublich schnell, wenn es darauf ankommt, fährt aber nicht unnötig riskant. So auch im neu konstruierten Mercedes Benz W 196 R, der damals vielen aufgrund seiner fließenden Linien der Karosserie merkwürdig erscheint. Aber die Idee, der Stromlinienform zu entsprechen, zahlt sich für die Konstrukteure in Verbindung mit Fangio mehr als aus. Diese Kombination demoralisiert die bis dahin die Rennstrecken beherrschenden Maseratis und Ferraris. Die Verbindung von Fangio und Mercedes-Benz, das perfekte Zusammenspiel von Fahrer, Fahrzeug und Rennstall, wird oft als nahezu magisch bezeichnet.

Der W 196 R ist ihm quasi auf den Leib geschneidert. Der Siegerwagen des Jahres 1954, Erster beim Grand Prix der Schweiz in Bern, in Monza, Reims und auf dem Nürburgring, ist ein lediglich 830 kg schwerer 2,5-Liter-Rennwagen, der von einem Reihen-Achtzylinder angetrieben wird. Erstmals verwendet man im W196 R eine Benzineinspritzung, die von Bosch entwickelt wurde. Der Motor liefert zum Entwicklungsendstand im Sommer 1955 rund 290 PS (213 kW), und beschleunigt den Monoposto auf über 275 km/h.

Obwohl er die Saison 1954 mit einem Maserati beginnt, u. a. weil der Mercedes-Benz erst Mitte des Jahres fertig wird, holt Juan Manuel Fangio im W 196 R in den Jahren 1954 und 1955 den Weltmeister-Titel.
Nach rund 40-jähriger Abstinenz vom Formel-1-Rennsport holen die Silberpfeile 1998 und 1999 erneut den Titel, Mika Häkkinen entscheidet die Weltmeisterschaften für sich. 2008 wird Lewis Hamilton mit McLaren-Mercedes der bis dahin jüngste Weltmeister in der Formel-1-Geschichte. In der Saison 2012 erzielt Nico Rosberg beim Großen Preis von China nach Juan Manuel Fangio in Monza 1955 den ersten Sieg eines Werks-Silberpfeils. Der Mythos Silberpfeil lebt und wird auch in Zukunft Geschichte schreiben. •

sent to the track had already earned the name "Silver Fish" for their aerodynamic silver body panelling. Mercedes-Benz and the Auto Union company, which was formed in 1932 by a merger of four car manufacturers (Audi, DKW, Horch and Wanderer), dominated Grand Prix racing up until 1939. With Rudolf Caracciola behind the wheel, Mercedes-Benz grabbed the European Championship title in 1935, 1937 and 1938—until the title victory was rendered unofficial by the outbreak of WWII in 1939. Time and again, top drivers from both racing teams like Nuvolari, Stuck, von Brauchitsch, Caracciola, Lang and Rosemeyer set new records in their Silver Arrows.

At the Reims race in July 1954, the Stuttgart team celebrated their first Grand Prix victory since 1939—the very first race with a new class of Silver Arrows. They had worked hard to develop a completely new racing car to meet the new 2.5-litre formula. Its futuristic, streamlined body featured a tubular space frame, and its direct fuel injection system set new technical standards.

This triumphant comeback also marked the start of a career for a man whose name would always be inseparably linked to Mercedes-Benz: Juan Manuel Fangio. He was a man who stood out from the "normal" circle of Formula 1 drivers: Cool, but not of a cold disposition. Quiet and considered, which may explain why he himself was held in such high consideration. He dominated Formula 1 racing in the 1950s. He was incredibly fast whenever necessary, yet his driving was never unnecessarily reckless. Fangio drove true to form in the newly constructed Mercedes-Benz W 196 R, which many at the time thought looked odd due to the flowing lines of its bodywork. But the designers' idea to streamline the shape more than paid off with Fangio behind the wheel. That combination of car and driver completely demoralized the Maseratis and Ferraris that had previously owned the race tracks.
Fangio and Mercedes-Benz formed a perfect symbiosis of driver, vehicle and racing team that was often spoken of as nearly magical.

The W 196 R was practically custom-made for Fangio. The most successful car of 1954, winning first place in the Swiss Grand Prix in Bern, in Monza, in Reims and at the Nürburgring, was a 2.5-litre racing car that weighed only 830 kg and was powered by a straight-eight engine. For the first time, the W 196 R featured a fuel injection system that had been developed by Bosch. At the final stage of development in the summer of 1955, its motor achieved 286 hp (213 kW), accelerating the single-seater to over 275 km/h.

Juan Manuel Fangio became world champion in the W 196 R in the years 1954 and 1955 – even though he started the1954 season in a Maserati because the Mercedes-Benz model was not ready until the middle of the year. After an almost 40-year hiatus from Formula 1 racing, the Silver Arrow conquered the title in 1998 and 1999 when Mika Häkkinen decided the world championship for himself. In 2008 Lewis Hamilton, driving for McLaren Mercedes, became the youngest Formula 1 world champion ever until that time. During the 2012 season Nico Rosberg won the Chinese Grand Prix, the first victory by a Silver Arrow factory model since Juan Manuel Fangio's 1955 win in Monza. The legend of the Silver Arrow lives on, and it will continue to write history. •

Beim Großen Preis von Europa 1954 ist es Juan Manuel Fangio mit seinem Silberpfeil, der das Rennen dominiert.

Juan Manuel Fangio dominates the 1954 European Grand Prix in his Silver Arrow.

Mechaniker der Rennabteilung tragen eine Stromlinienkarosserie des Mercedes-Benz Formel-1-Rennwagens W 196 R aus dem Jahr 1954.
Mechanics from the racing division carry the streamlined body of the Mercedes-Benz W 196 R Formula One racing car from 1954.

[redacted]

Versuchs-Chef Rudolf Uhlenhaut und Technischer Direktor Prof. Dr. Fritz Nallinger vor dem Mercedes-Benz Formel-1-Rennwagen W 196 R.
Head of testing Rudolf Uhlenhaut and technical director Prof. Dr. Fritz Nallinger in front of the Mercedes-Benz W 196 R Formula One racing car.

Ein junger Seifenkistenfahrer am Steuerrad des Mercedes-Benz Formel-1-Rennwagens W 196 R . Daneben Hans Herrmann, 3. beim Großen Preis der Schweiz (1954).
A young soapbox driver behind the wheel of the Mercedes-Benz W 196 R Formula One racing car. With him: Hans Herrmann, who finished 3rd in the 1954 Swiss Grand Prix.

Die argentinische Formel-1-Legende Juan Manuel Fangio beim Großen Preis von Italien 1954.
Argentinian Formula One legend Juan Manuel Fangio at the 1954 Italian Grand Prix.

JUAN MANUEL FANGIO

Der argentinische Rennfahrer Juan Manuel Fangio gilt bis heute als einer der besten Formel-1-Fahrer. Und obwohl man ihm nachsagte, dass er einen eher unspektakulären Fahrstil pflege, sprachen die Stoppuhren fast immer für ihn. Der „Maestro" fuhr schonend - sehr zur Freude der Monteure, wenn sie seine Motoren oder Getriebe überprüften - was den großen Vorteil hatte, dass er selten im Rennen ausfiel. Vielleicht auch, weil er Mechaniker gelernt hatte und daher das nötige Feingefühl für die Technik mitbrachte. Mit den Rennwagen, die er fuhr, schien er eine nahezu magische Verbindung zu haben. Und die zeigte sich vor allem in den Mercedes-Benz Silberpfeilen der 1950er Jahre.

To this day, Juan Manuel Fangio of Argentina is considered one of the very best Formula 1 racing drivers. Although his driving style was said to be unspectacular, the stopwatch was almost always in his favour. The "maestro" drove smoothly, much to the delight of his crew whenever they checked his engines or transmissions – the great advantage being that he seldom had to fall out of a race. Perhaps that was also due to his training as a mechanic and to the keen sense of engineering that brought with it. He was said to have had an almost magical connection to the racing cars he drove. This was most apparent with the Mercedes-Benz Silver Arrows of the 1950s.

Die Silberpfeile stachen nicht nur visuell aus dem Fahrerfeld heraus, auch ihre Leistungen waren beeindruckend.

The Silver Arrows didn't just stand out due to their appearance - their performance was impressive too.

12

W196

ÜBER DEN WERT DES UNWIEDERBRINGLICHEN.

ON THE VALUE OF THE IRRETRIEVABLE.

Unermesslich wertvoll: Der W196 von Juan Manuel Fangio besitzt als Stilikone und einstige Galionsfigur der Silberpfeile einen historischen, kulturellen und emotionalen Wert, der in Zahlen nicht greifbar ist.
Utterly priceless: Juan Manuel Fangio's W 196 is a true icon of style and a historic figurehead amongst Silver Arrows; its historical, cultural and emotional value that cannot be reflected in numbers.

Der Preis eines Objektes ist kein Wert an sich. Aber er ist Ausdruck der öffentlichen Wertschätzung und Begehrlichkeit, die einem Objekt entgegengebracht wird. Der erzielbare Preis kann niemals ein Grund dafür sein, ein einmaliges historisches Automobil – unbestreitbares Zeugnis der technischen Entwicklung einer Epoche – zu erhalten und zu bewahren. Es ist die Einzigartigkeit, die Unwiederbringlichkeit, die durch nichts zu ersetzende Rarität eines Objektes, die dazu führt, dass einzelne Automobile Preise erzielen, die vergleichbar sind mit den besten Werken herausragender Künstler. Sie lassen einen Markt entstehen, der nicht Selbstzweck ist sondern Ausdruck dieser Wertschätzung.

Während praktisch jede Maschine mit viel Aufwand in einen Zustand versetzt werden kann, der zum Zeitpunkt ihrer Auslieferung dem Original entspricht, kann keinAufwand dieser Welt einem Automobil die wahren Spuren seiner Vergangenheit und damit seine eigene Identität und Geschichte wiedergeben, nachdem sie einmal zerstört oder durch eine Restauration genommen wurden.

Aus diesem Grund ist der Mercedes-Benz W 196 R Grand-Prix-Rennwagen, der im Juni 2013 bei der Bonhams Auktion in Goodwood versteigert wurde, ein einzigartiges Objekt der Zeitgeschichte, das authentisch Zeugnis ablegt von den legendären Schlachten der besten Fahrer ihrer Zeit auf den Grand-Prix-Strecken dieser Welt: Juan Manuel Fangio gewann mit ihm 1954 den Großen Preis von Deutschland auf dem Nürburgring und den Großen Preis der Schweiz in Bern – gleichzeitig die ersten Grand-Prix-Siege des neuen Monoposto von Mercedes-Benz nach dem Zweiten Weltkrieg überhaupt. Und in genau diesem Zustand ist der W 196 R bis heute erhalten. Ohne Zweifel ist er damit eines der geschichtsträchtigsten Automobile des 20. Jahrhunderts und eines der bedeutendsten Fahrzeuge aller Zeiten. Seine Rarität besteht in der Einzigartigkeit und Unwiederrufbarkeit der Spuren, die diese Geschichte auf ihm hinterlassen hat. Und die ihn, trotz allem, als funktionierendes Zeugnis des Strebens nach technischen Höchstleistungen und Erfolg die Zeit überdauern ließ. Aus gutem Grund ist er damit das teuerste, jemals bei einer Auktion versteigerte Fahrzeug.

Welcher Ort auf der Welt wäre für eine solche Auktion besser geeignet gewesen als Goodwood House? Einmal im Jahr verwandelt Lord March seinen riesigen Vorgarten in eine Bühne, auf der hunderttausende Besucher des „Good-wood Festival of Speed“ deutlich machen, dass Wertschätzung und Begeisterung für ungewöhnliche Zeugnisse automobiler Entwicklung keine elitären Angelegenheiten sind.

Während Auktionator Robert Brooks bei vorherigen Lots zurückhaltende Bieter schon mal mit 500-Pfund-Schritten aus der Reserve locken muss, herrscht im Zelt absolute Ruhe, als sich die Bieter beim Mercedes-Benz Silberpfeil vor Ort und am Telefon in 500.000 Pfund Schritten auf schwindelerregende 17,5 Millionen Pfund schrauben. Mit allen Gebühren, die der Käufer zu tragen hat, liegt der Preis letztendlich bei atemberaubenden 19,6 Millionen Pfund (22,7 Mio. Euro bzw. 31,6 Mio. US-Dollar). Der höchste, jemals bei einer Auktion erzielte Preis für ein Automobil überhaupt – vor allem aber ein Zeugnis der Wertschätzung für die Unwiederbringlichkeit und Rarität der Geschichte eines einzigartigen Fahrzeugs. •

The price that an object commands is not its true value, although it does reflect the public appreciation and desirability which that object is shown. The cost ascribed to a one-of-a-kind historic car is never reason enough to own and maintain it – such cars are indisputably a testimony to the technical developments of their time. Instead, the attributes that makes something unique, irrevocable, irreplaceably rare are why individual cars can attain prices which are comparable to the finest works of outstanding artists. And these cars can create a market which is not an end unto itself, but rather an expression of how much they are held in high regard.

With enough effort, practically any car can be returned to a condition that corresponds to its original state when it left the factory. But there is no effort of any kind that can replace the original traces of a car‘s history, which would bring back its own distinct identity and individual history if it is damaged or lost due to restoration.

This is why the Mercedes-Benz W 196 R Grand Prix race car auctioned off at Bonham‘s in Goodwood in July 2013 is a unique object. It is an authentic witness to the history of the epoch‘s best drivers and the legendary races they drove at Grand Prix sites around the world. In 1954, Juan Manuel Fangio used it to win the German Grand Prix at Nürburgring and the Swiss Grand Prix in Bern, which were also the first post-WWII Grand Prix victories of the new Mercedes-Benz single-seater. And the W 196 R has stayed in that very condition up to this day. This indubitably makes it one of most historically important cars of the 20th century as well as one of the most significant vehicles of all time. Its iconic personality has its roots in the unique, exclusive and irrevocable marks that history left upon it. And nevertheless, despite everything it also stood the test of time as living proof of the quest to achieve the utmost in technical innovation and success. All valid reasons why this was the most expensive car ever to be auctioned.

Is there a place anywhere in the world that would have been better suited to auction off this car than Goodwood House? Once a year, Lord March opens up the grounds of his home – admittedly no modest affair – to host the “Goodwood Festival of Speed”. With hundreds of thousands of visitors in attendance, the Festival clearly shows that the appreciation and enthusiasm people feel for unusual items from automotive history is by no means an elite affair.

During the first several lots, auctioneer Robert Brooks had to nudge hesitant bidders along at increments of £500, but there was tense calm in the tent when the Mercedes-Benz Silver Arrow came under the hammer. Both on site and over the phone, the bid advanced at awe-inspiring intervals of £500,000, sending the price skyrocketing to £17.5 million. After all of the fees and commissions, the final price amounted to a breathtaking £19.6 million or €22,700,000 or $31,600,000.00. This was the highest price a car ever fetched at auction – and above all, confirmation of how much people care about the irretrievably rare character of a unique car‘s history. •

Das Lenkrad des W 196 R, von den Spuren der Vergangenheit gezeichnet, ist stummer Zeuge einer großen Epoche des Motorsports.
The steering wheel of the W 196 R is marked by the traces of history, bearing silent witness to a great period in motorsport.

12

12

Ein besonders auffallendes Merkmal des W 196 R ist eine nach außen gewandte aber dennoch der stromlinienförmigen Karosserie angepasste Auspuffanlage.
A particularly striking feature of the W 196 R is its exhaust system, which protrudes from the car but is shaped to fit in with the streamlined body.

[redacted]

Das technische Innenleben des W 196 R wirkt heute sehr puristisch und übersichtlich, war damals jedoch das ausgereifteste, was man sich vorstellen konnte.
Technically, the interior of the W 196 R looks very purist and straightforward. At the time, it was the most sophisticated design imaginable

Für 22,7 Mio. Euro wurde Fangios W 196 R im Juni 2013 auf dem Goodwood Festival of Speed versteigert.
Fangio's W 196 R was auctioned off for €22.7 million in June 2013 at the Goodwood Festival of Speed.

DER WEGGEFÄHRTE
THE PERFECT PARTNER

KLAUS KAISER ÜBER DIE GEMEINSAME ZEIT MIT EUGEN BÖHRINGER. EIN RÜCKBLICK.

A TRIBUTE TO EUGEN BÖHRINGER. KLAUS KAISER REFLECTS.

Eugen Böhringer (rechts), das Damenteam Ewy Rosqvist / Ursula Wirth (3. Platz) und Juan Manuel Bordeu (4. Platz) bei der Siegerehrung zum Großen Straßenpreis von Argentinien, 1963 in Buenos Aires.
Eugen Böhringer (right), female duo Ewy Rosqvist and Ursula Wirth (3rd place) and Juan Manuel Bordeu (4th place) at the award ceremony in Buenos Aires for Rally Argentina 1963.

[redacted]

Eugen Böhringer, 1922 in Rotenberg bei Stuttgart geboren, zählte zu den ganz großen Mercedes-Benz Werksfahrern. Einer der für das Rennfahren geboren zu sein schien, ein herausragender Rallyepilot und absolutes Ausnahmetalent. Schon in jungen Jahren entwickelte er eine ausgeprägte Leidenschaft für Autos und Rennen, ein Grund weshalb er unbedingt Automobilmechaniker werden wollte. Die Realität sah jedoch zunächst ganz anders aus. Weil er das Hotelrestaurant seiner Eltern übernehmen sollte, absolvierte er eine Ausbildung zum Koch. In seiner Freizeit widmete er sich jedoch zunehmend seiner wahren Leidenschaft, dem Rennfahren. Zum professionellen Rennsport gelangte er dann schließlich. Im Alter von 36 Jahren kam er schließlich durch eine Wette mit Freunden zum Profi-Rennsport. Das vereinbarte Ziel: das Geschicklichkeitsturnier des Stuttgart-Untertürkheimer Motorsportclubs auf dem Cannstatter Wasen gewinnen.
Eugen Böhringer trat mit seinem privaten Mercedes-Benz 219 (W 105) an und gewann souverän. Daraufhin folgten zahlreiche Siege im Breitensport und schließlich der Wechsel ins Profilager durch ein Angebot, das er nicht abschlagen konnte: eine Werkfahrerstelle bei Mercedes-Benz, die zahlreiche Rallyeerfolge nach sich zog, u. a. den Rallye-Europameistertitel 1962, den er mit der Mercedes-Benz 230 SL „Pagode" einfuhr. Bis zu seinem Tod im Juni 2013 wurde Böhringer oft im selben Atemzug mit der „Pagode" genannt. Sein langjähriger Weggefährte und Co-Pilot Klaus Kaiser erinnert sich an seine schnellen Zeiten mit einer der schillernden Persönlichkeiten der Mercedes-Benz-Rennsport-Geschichte.

„Erfolge hatten wir viele, aber kein Erfolg für einen Fahrer ohne den Beifahrer", Klaus Kaiser bringt seine besondere Beziehung als Co zu Eugen Böhringer, typisch schwäbisch kurz und knapp auf den Punkt. Er macht keinen Hehl daraus, dass sie nicht die besten Freunde waren, dafür waren sie ein perfekt eingespieltes Team, mit gegenseitigem Respekt vor Person und Leistung. Und die wird bei anspruchsvollen Rallyes Fahrer und Beifahrer abverlangt, denn ohne das perfekte Timing des Co-Piloten ist im Rallye-Motorsport ein großer Erfolg so gut wie unmöglich. Wie ein präzises Uhrwerk müssen beide Seiten funktionieren. Während der Co-Pilot aus dem sogenannten „Gebetsbuch" vorliest, wann welche Kurve, Kreuzung, Abzweigung oder Piste kommt, muss der Fahrer reagieren und sich zu 100 Prozent auf die Ansagen seines Beifahrers verlassen. Und das konnte Böhringer, denn mit Klaus Kaiser hatte er den perfekten Co-Piloten.

Angefangen hatte ihr gemeinsamer Weg beim Training zur Tulpen-Rallye 1963, im Elsass. Für die Rallye, die nachts stattfand, wurde offiziell „eigentlich nur tagsüber trainiert". Doch das war Böhringer nicht genug. Kaiser erkannte damals sofort den Ehrgeiz des Stuttgarter Rennfahrers. Und auch seine Chance: *„Nach dem Nachtessen sagte Böhringer plötzlich zu mir: Jetzt fahren wir noch raus. Die Strecke muss man bei Nacht durchgehen, da sieht sie ganz anders aus, als am Tag. Wir nutzten die Zeit und die Verhältnisse bei Nacht, um den Aufschrieb im Gebetsbuch zu verbessern. Und auf dem Rückweg sagte Böhringer zu mir: „Jetzt kannst du fahren"*. Das tat Kaiser auch und bewies Böhringer, dass er selbst auch gut am Steuer war. *„Ich war schließlich Fahrlehrer."* Das war der Anfang einer erfolgreichen Rennfahrerverbindung, die oft auf die Probe gestellt wurde. Die oft auf die Probe gestellt wurde. Denn die

Eugen Böhringer, born in Rotenberg just outside Stuttgart in 1922, was one of Mercedes-Benz's most famous factory drivers – an exceptionally talented rally driver who seemed born to race. Passionate about cars and racing from a young age, he wanted to become a car mechanic, but things turned out differently and he ended up training as a cook, as his parents wanted him to take over their hotel and restaurant. However, Böhringer devoted more and more of his free time to his true calling of racing, finally entering the professional motorsport world at the age of 36 after making a bet with friends that he could win a local tournament organised by the Stuttgart-Untertürkheim Motorsport Club.

Böhringer competed in his own car, a Mercedes-Benz 219 (W 105), and won the race easily. He followed this up with more victories in amateur events before becoming a professional when Mercedes-Benz made him an offer he couldn't refuse: a position as a factory driver. This opened the door to a series of rally wins, including the 1962 European Rally Championship, in which Böhringer raced to victory in the "Pagoda" – the Mercedes-Benz 230 SL. Böhringer, who passed away in June 2013, was often mentioned in the same breath as the Pagoda. Klaus Kaiser, his former co-driver, recalls the thrilling times he spent in the cockpit with one of the biggest names in Mercedes-Benz racing history.

"We had a lot of wins, but a driver can't win without the co-driver", comments Kaiser in typically terse Swabian style. He makes no secret of the fact that he and Böhringer weren't the best of friends, but proudly points to their well-oiled partnership built on mutual respect for each other as people and professionals. It's this partnership that counts most of all in challenging rallies when both the driver and co-driver are put to the test – without the co-driver's perfect timing, major success in rallying is as good as impossible. The driver and co-driver have to work together as one: the co-driver reads from the pacenotes and informs the driver of coming bends, junctions, turnings and dirt roads, and the driver has to react accordingly and be able to completely trust everything the co-driver tells them. Böhringer could: in Klaus Kaiser, he had the perfect co-driver.

The two first teamed up for the 1963 Tulip Rally in Alsace. Although the rally was to be held at night, the official training periods were scheduled for during the day. For Böhringer, this was insufficient. Kaiser immediately recognised Böhringer's ambition and the opportunity they had. "After we finished our evening meal, Eugen suddenly said to me 'Let's go back out again. We need to drive the course by night, it'll look very different than it does in the daytime.'" remembers Kaiser. "So we familiarised ourselves with the night-time conditions and improved our pacenotes. On the way back, Böhringer said to me: 'Now, you drive!'". So Kaiser took the wheel and showed Böhringer that he was a good driver too. "I was a driving teacher, after all." This was the start of a successful partnership, one that would be put through its paces in rally after rally. Races then were less closely regulated than they are today and were often tough, tense affairs.

"Böhringer was always good for a surprise", says Kaiser about his colleague. "One day he came over to me and asked 'Do you have time?'". I replied "'Time for what?', and he told me he needed a driver for a rally in Sweden. So I agreed." Böhringer was looking for a reliable

VII. Großer Straßenpreis von Argentinien für Tourenwagen, 1963. Eugen Böhringer und Klaus Kaiser mit ihrem Mercedes-Benz 300 SE (W 112) kommen als Sieger ins Ziel.
Eugen Böhringer and Klaus Kaiser win the 7th Rally Argentina in their Mercedes-Benz 300 SE (W 112) in 1963.

noch nicht so stark wie heute reglementierten Rennen von damals versprachen Nervenkitzel pur. *„Böhringer war immer für eine Überraschung gut"*, beschreibt Kaiser den Rallye-Fahrer. *„Da kam er doch einfach zu mir und fragte: Hast Du Zeit?"*. Ich erwiderte damals *„Was hab ich Zeit?, er sagte er brauche einen Fahrer für Schweden. Ich sagte Okay."* Böhringer brauchte einen verlässlichen Co-denken. Klaus Kaiser schien als strenger, aufmerksamer Fahrlehrer alle Kriterien zu erfüllen, dennoch wollte es Eugen Böhringer genau wissen. Eine Trainingstour in der hügeligen Gegend um Stuttgart stand auf dem Programm. *„Böhringer hat mir erklärt, dass die Strecke in Schweden nicht bekannt sei und erst acht Stunden vor dem Rennen Preis gegeben wird."* Das wollte er damals mit Klaus Kaiser trainieren. *„Da sind wir ins Umland von Schnait nach Schornbach hoch und er hat mir erklärt, was ich zu notieren habe. Auffallende Dinge wie einen Telegrafenmast, ein besonderes Haus an der Strecke. Es sollte nur wirklich auffallende Dinge notieren, die als Anhaltspunkte dienten. Dann hat er mir den Charakter von Kurven diktiert. Auf dem Rückweg hat er gesagt, jetzt umgekehrt, jetzt du."* Böhringer war ein besonderer Fahrer, der viel Wert auf Geschick, Reaktion und Durchhaltevermögen legte. Mit seiner Wahl von Klaus Kaiser als Co-Piloten erwies er zudem ein gutes Gespür für kompetente Leute. Schweden war die erste erfolgreiche, gemeinsam gemeisterte Fahrt im Mercedes-Benz 220 SE (W 111). *„Teilweise im Schnee und Eis, obwohl es Sommer war"*; erinnert sich Kaiser. *„Das war hart."*

„Früher war das überhaupt anders als heute, da fuhr man nicht nur ein paar Stunden", weiß Kaiser. *„Zum Beispiel bei der Rallye Spa-Sofia-Liège 1963, da waren wir eine lange Strecke am Stück unterwegs, 1500 Kilometer."* Große Unterhaltungen gab es kaum, denn mit den Helmen, dem anfälligen Funk und dem ohrenbetäubendem Lärm beim Fahren war die Kommunikation oft auf das Wesentliche beschränkt. *„Ich las aus dem Gebetsbuch vor, Eugen Böhringer reagierte, hakte hin und wieder nach, wenn etwas unklar war. Zeit für längere persönliche Gespräche gab es während der Rennen allerdings nicht."* In solchen reaktionsschnellen Rennsituationen passieren auch Fehler. *„Die dürfen eigentlich nicht passieren, tun sie aber. Ein Verwechseln von Links oder Rechts war zwar im Allgemeinen erlaubt. Das kam aber zum Glück nur selten vor."*
Kommunikation ist alles. Gerade im Motorsport. Kurze und knappe Anweisungen des Beifahrers aus dem Gebetsbuch und die fahrtechnische Reaktion des Rennpiloten sind die Grundlage eines jeden Erfolgs, und man entwickelt als Team trotz Standardkommandos eine ganz individuelle Kommunikation. *„Das muss man auch, man braucht als Team im Motorsport seine eigene Sprache, damit man in Sekunden reagieren kann"*, erklärt Kaiser. Gerade auf der „Lüttich" kam es zum Härtetest. Kaiser erinnert sich *„Ich habe zwölf Stunden lang nur vorgelesen, nachts mit Lampe und das Buch im Schoss. Eugen Böhringer hat reagiert."* Eine Situation, die keinen Raum für private Gespräche ließ. *„Bemerkenswert ist, dass man es heute nicht anders macht als vor 40, 50 Jahren, nur dass es heute elektronische Hilfs-*

co-driver who could think ahead, and Kaiser – a precise and attentive driving teacher – seemed to tick every box. But Böhringer wanted to be sure Kaiser was the right choice, so the two went for a practice drive in the hills outside Stuttgart. "Eugen told me that the route in Sweden was secret and would only be announced eight hours before the start of the rally", explains Kaiser. "So we drove into the countryside around Schnait, up to Schornbach, and he told me what I needed to pay attention to. Memorable things like a telegraph pole, a certain house by the road – it had to be really striking things we could note. Then he dictated the character of the bends to me. On the way back, he said 'Now, let's switch – you take the wheel.'"
Böhringer was a remarkable driver who stood out for his skill, rapid reaction and stamina. In his choice of Klaus Kaiser as co-driver, he also demonstrated a good nose for a partner with the right abilities. The 1962 Midnight Sun Rally in Sweden was the pair's first successful race in the Mercedes-Benz 220 SE (W 111). "Part of the rally was on snowy and icy roads, even though it was summer", remembers Kaiser. "That was hard."

"Back then, things were totally different to today – you didn't just drive for a couple of hours at a time", Kaiser continues. "During the 1963 Spa-Sofia-Liège Rally, for instance, we once drove for 1500 kilometres without stopping." There wasn't much in the way of in-car entertainment then either – in fact, communication between driver and co-driver was often kept to a minimum due to their helmets, the fragility of the radio system and the overwhelmingly loud driving noise. "I read from the pacenotes and Eugen reacted, asking me questions from time to time when something was unclear, but there was no time for long personal conversations during the rally." In race situations where rapid reactions are crucial, mistakes can also happen easily. "They shouldn't, but they do", Kaiser remarks. "Mixing up left and right isn't forbidden, but fortunately it rarely happened."
Communication is everything – especially in motorsport. Clear, concise directions from the co-driver followed by prompt, skilled reactions by the driver are key to every victory. So despite there being standard commands, teams develop their own individual communication systems. "You have to do that – as a motorsport team, you need to have your own language so that you can react in seconds", states Kaiser. The Liege rally put this to the test. "I just read the pacenotes for twelve hours", he recalls. "This was at night, so I had the book in my lap and read it with a lamp. I read, Eugen reacted." Not exactly the perfect atmosphere for personal conversation. "What's notable is that today, nothing about this has changed. It's the same as it was 40-50 years ago, although one big difference is the introduction of electronic tools like cruise control. Back then, all we had were mechanics who we communicated with by radio, and who weren't always that precise." Almost everything was down to Böhringer and his co-driver.

"Eugen was never very talkative when we were sitting in the cockpit", remembers Kaiser. "But over time, you get to know one another well." Although their relationship was

mittel wie den Speedpiloten gibt. Damals gab es nur Mechaniker, die über Funk kommunizierten, die waren nicht immer so genau.“ Es lag also alles an Böhringer und seinem Beifahrer.

„Wenn wir im Cockpit saßen, war Eugen Böhringer nicht ganz so gesprächig“, erinnert sich Kaiser. *„Aber man lernt sich trotzdem mit der Zeit ganz gut kennen.“* Trotz eines eher professionellen statt tief freundschaftlichem Verhältnisses, hatten sich beide gut verstanden. *„Man baut, das ist ganz wichtig für ein Rallye-Team, ein großes Vertrauen auf.“* Kaiser erinnert sich an die gemeinsamen Anfänge in den 1960er Jahren in Schweden: *„Eugen Böhringer wusste immer ganz genau was er wollte. Zu mir sagte er damals: Du bist mein nächster Copilot.“* Sein Siegeswillen war legendär, aber auch sein nicht überhitztes Temperament: *„Dafür hatte er einen unglaublichen Ehrgeiz, eine Eigenschaft, die bei vielen heute nicht mehr so verbreitet ist.“* Böhringer wollte siegen, ganz nach vorne, aber nicht um jeden Preis. Dafür hatte er mit Klaus Kaiser seinen idealen Co-Piloten gefunden und das Rallyefahren im Team perfektioniert. Vier Jahre lang fuhren sie gemeinsam viele Rennen und eroberten zahlreiche erste Plätze.

Am legendärsten war wohl die Rallye Spa-Sofia-Liège 1963, in dem sie mit der roten „Rennpagode“ 230 SL den Sieg einfuhren. Die rote „Pagode“ ist es auch, mit der Eugen Böhringer immer noch am häufigsten in Verbindung gebracht wird. Sie ist sein „Markenzeichen“. •

always more a professional one than any kind of deep friendship, the two understood each other well. “You build a great amount of trust to each other, and that’s very important for a rally team”, comments Kaiser. “Eugen always knew exactly what he wanted. Back then, he said to me ‘You’re my next co-driver.” Böhringer’s thirst for victory was legendary, as was his very level temperament. “But he had unbelievable ambition, something that’s a lot rarer in today’s rally world”, Kaiser says. Böhringer wanted to win, to be in pole position, but not at any price. And in Klaus Kaiser, he’d found the perfect co-driver with whom he could perfect his rallying technique. The two raced together for four years, notching up an impressive tally of first places.

Most legendary of all was the 1963 Spa-Sofia-Liège Rally, won by Böhringer and Kaiser in a red 230 SL “Racing Pagoda”. It’s this red Pagoda – his trademark – that is still most commonly associated with Eugen Böhringer today. •

IVB-4052

GREEN GIANT

AUF WELTTOURNEE
ON A WORLD TOUR

DER LEGENDÄRE PRINZ-HEINRICH-WAGEN ZU GAST AUF DEM CONCOURS D'ELÉGANCE IN PEBBLE BEACH.

THE LEGENDARY PRINZ HEINRICH BENZ AT THE CONCOURS D'ELEGANCE IN PEBBLE BEACH.

Männer in dicken Lederjacken, Staubkappen und Brillen donnern auf hubraumgewaltigen Monstern über öffentliche Straßen, ach was, staubige Überlandwege, die zu dieser Zeit noch von Postkutschen beherrscht werden. Bedrängt nur von Handkarren und aufgescheuchten Hühnern, die die Straßen für sich beanspruchen. Die Rennen sind lang, finden bevorzugt zwischen den großen europäischen Metropolen Paris, Madrid, Berlin, Wien statt, und die Regeln sind einfach: derjenige, dessen Auto durchhält und als erster ankommt, hat gewonnen. Nach diesem einfachen Muster begeistern Autorennen zu Beginn des 20. Jahrhunderts in vielen Ländern Europas die Massen. Der Höhepunkt sind die Rennen um den „Gordon Bennett Cup", die der amerikanische Zeitungsverleger James Gordon Bennett zwischen 1900 und 1905 ausgeschrieben hatte. Und die wilde Jagd über den Kontinent wäre wohl noch über viele Jahre weitergegangen, wenn nicht das Langstreckenrennen Paris – Madrid 1903 in Bordeaux ein tragisches Ende genommen hätte. Bei dem Versuch, dem allgegenwärtigen Chaos an der Strecke auszuweichen, schießen einige der Teilnehmer mit hoher Geschwindigkeit über die Straßen hinaus, bleiben an Alleebäumen hängen oder geraten beim Aufprall auf andere Hindernisse in Brand. Gerüchte von Hunderten von Verletzten und Dutzenden von Toten machen die Runde. Das Rennen kostet dann doch „nur" drei Zuschauer und fünf Teilnehmer das Leben. Aber das Ende der wilden Rennerei auf öffentlichen Straßen ist damit besiegelt. In den belgischen Ardennen nahe des Städtchens Bastogne entsteht stattdessen der erste geschlossene Motorsport-Kurs, der „Circuit des Ardennes", dem bis in die Zwanzigerjahre weitere europäische Strecken folgen.

„DAS RENNEN KOSTETE DANN DOCH „NUR" DREI ZUSCHAUERN UND FÜNF TEILNEHMERN DAS LEBEN."

In Deutschland stehen Zuverlässigkeitsfahrten hoch im Kurs. Sie sollen die Alltagstauglichkeit des Automobils im sportlichen Wettbewerb demonstrieren. 1903 ruft der Maler und Automobilist Hubert von Herkomer die erste derartige Veranstaltung ins Leben, die von 1905 bis 1907 jährlich ausgetragen wird. 1907 hebt Heinrich von Preußen, der automobilbegeisterte Bruder des deutschen Kaisers, als Ersatz für die erfolgreichen Herkomer-Fahrten eine eigene Tourenwagen-Konkurrenz aus der Taufe: die Prinz-Heinrich-Fahrt. Ausdrücklich nicht für Renn- sondern für viersitzige Vier- oder Sechszylinder Tourenwagen, die zur Fahrt auf öffentlichen Straßen zugelassen sein mussten. Aber wenn man ehrlich ist, hatten sich gegenüber den wilden Langstrecken-Rennen der Jahrhundertwende, lediglich ein paar Regularien geändert - Durchhaltevermögen über lange Strecken bei hohen Geschwindigkeiten galt nach wie vor als größte Herausforderung für die frühen Automobile. Und Mut als die wichtigste Tugend ihrer Chauffeure. Vor allem, wenn sie immer wieder zwischen den bis zu 350 Kilometer langen Tagesetappen das Potential ihrer Wagen in Rennen auf abgesperrten Strecken unter Beweis zu stellen hatten. Entspannt waren diese Renneinlagen nur für die Kontrolleure des Veranstalters, die ne-

Men in thick leather jackets, caps and goggles driving monsters with powerful engines thunder along open roads, or rather dusty lanes that back in those days were still dominated by stagecoaches. The only things in their path are pushcarts and a few startled chickens that are hogging the road. The races are long, link up major European cities such as Paris, Madrid, Berlin and Vienna, and follow simple rules: the car that holds out longest and comes in first wins. Based on this simple formula, motor racing drew enthusiastic crowds all over Europe at the beginning of the 20th century. The highlight of the racing year were the "Gordon Bennett Cup" races organised by the American newspaper publisher James Gordon Bennett between 1900 and 1905. This wild chase across the continent would probably have continued for many years if the Paris to Madrid race in 1903 had not come to a tragic end in Bordeaux. Attempting to avoid the chaos all along the route, several participants shot off the road at high speed, crashing into trees or striking other obstacles and bursting into flames. Hundreds of people were said to be wounded and dozens dead. It turned out that "only" three spectators and five participants had lost their lives. But it marked the end of open road racing. Instead, the first closed motorsport track, the "Circuit des Ardennes", was built near the small town of Bastogne in the Belgian Ardennes, with other European circuits following in the years leading up to the 1920s.

"IT TURNED OUT THAT 'ONLY' THREE SPECTATORS AND FIVE PARTICIPANTS LOST THEIR LIVES."

In Germany, road endurance tests grew in popularity. These were sporting contests that aimed to demonstrate the road capability of cars. In 1903, the painter and motorist Hubert von Herkomer created the first event of this kind, which was held annually from 1905 to 1907. In 1907, Prince Heinrich of Prussia, brother of the German Kaiser and a motor car enthusiast, launched his own competition for touring cars, called the Prinz Heinrich Tour, which followed on from the successful Herkomer races. It only admitted four-seater touring cars with four or six cylinders that had to be authorised to drive on public roads. But essentially, compared with the crazy long-distance races of the turn of the century, only a few formalities had changed: holding out long distances at high speeds was still the greatest challenge for the first motor racing cars, and courage the most important virtue for their drivers. Especially when, as was often the case, in between the daily legs of the race – which were up to 220 miles long – they had to prove their car's capability in short sprints on closed sections of road. Only the organiser's controllers, who rode fearlessly on the back seat of each car throughout the race as passengers behind the driver and co-driver, were given a rest, as they were allowed to leave the car during these short interims.

The 1910 rally went from Berlin to Homburg vor der Höhe, via Brunswick, Kassel, Nuremberg, Strasbourg and Metz – a distance of 1,208 miles. Benz & Cie., winner of the first Prinz Heinrich Tour in 1908, naturally wanted to demonstrate the performance of its automobiles at what was Europe's most important

ben Chauffeur und Beifahrer als Passagier auf der gesamten Strecke in jedem Fahrzeug furchtlos auf dem Rücksitz mitfuhren: sie durften kurzzeitig von Bord gehen.

1910 ging die Fahrt über 1.945 km von Berlin über Braunschweig, Kassel, Nürnberg, Straßburg, Metz nach Homburg v.d.H. Natürlich wollte auch Benz & Cie., Gewinner der ersten Prinz-Heinrich-Fahrt von 1908, die Leistungsfähigkeit seiner Produkte bei der damals bedeutendsten Motorsportveranstaltung Europas unter Beweis stellen. Eigens dafür wurden deshalb zehn Wagen mit aerodynamisch optimierter Karosserie und dem dafür charakteristischen Spitzheck konstruiert. Vier davon mit 5,7 l und sechs mit 7,3 l Hubraum. Pioniergeist und Innovationen waren schon damals der Treibstoff der noch jungen Automobilindustrie. Und so rüstete Benz & Cie seine Wagen mit hoch effizienten Motoren aus, die über Doppelzündung und Vierventiltechnik verfügten. In dieser Kombination waren die Prinz-Heinrich-Wagen ihrer Zeit technisch weit voraus – auch wenn sie den Wettbewerb, für den sie ursprünglich konstruiert worden waren, nicht gewinnen konnten.

Es grenzt an ein Wunder, dass zwei dieser Automobile die mehr als hundert Jahre seit dem Rennen überlebt haben. Das Exemplar mit der Nummer „38" besitzt selbst noch den modifizierten Motor mit 5,4 Litern Hubraum, der ihm für seinen Einsatz bei der Zar-Nikolaus-Tourenfahrt im selben Jahr eingebaut worden war.

Nach einer aufwändigen, originalgetreuen Restauration, die die Experten von Mercedes-Benz Classic mit vielen über die ganze Welt verstreuten Spezialisten organisiert hatten, legt der dunkelgrüne Wagen nun auf den wichtigsten Klassiker-Veranstaltungen der Welt Zeugnis vom Pionier- und Innovationsgeist des Hauses ab. Klar, dass er im Jahr seiner Präsentation auch bei der Veranstaltung nicht fehlen durfte, die für viele zu den bedeutendsten des Jahres zählt: Pebble Beach.

Die Klassik-Veranstaltung zählt weltweit zu den renommiertesten Treffen historischer Automobile, und der Concours d'Elégance markiert den Höhepunkt eines Wochenendes, das vollständig im Zeichen eleganter, feiner und herausragender Automobile steht. Mercedes-Benz und die hochkarätige Veranstaltung, die traditionell am dritten August-Wochenende an der amerikanischen Pazifikküste stattfindet, verbindet eine erfolgreiche Tradition: Zur langen Liste der Auszeichnungen zählen seit 1950 mehrere Gesamtsiege sowie mehr als 120 Klassensiege und Spezialpreise. Im Jahr zuvor hatte ein Mercedes-Benz Typ S mit Saoutchik-Karosserie die begehrte Auszeichnung „Best of Show" gewonnen.

2013 war der Mercedes-Auftritt von ganz anderer Natur. Der kalifornische Küstennebel, der zu Pebble Beach gehört, wie die Panama-Hüte der Juroren, legt ein mildes Licht über die zahllosen Preziosen am 18. Loch. Zähmt die überflutende Formensprache und den oftmals großzügigen Einsatz von Lack und Chrom vieler Karossen, neben denen der Prinz-Heinrich-Wagen in seiner formalen Schlichtheit fast unbemerkt steht. Seine frische Lackierung tut ein übriges, bildet mit dem perfekt geschnittenen Rasen eine sanfte Melange in Grün, die nur Kenner zu entschlüsseln wissen – und macht den frisch restaurierten grünen Giganten so doch zum heimlichen Höhepunkt einer der bedeutendsten Klassiker-Veranstaltungen des Jahres. •

motor sport event at the time. Ten aerodynamic cars with the characteristic tapered rear were built especially for the race. Four had 5.7-litre engines, while the other six had engines with a displacement of 7.3 litres. Pioneering spirit and innovation were already the driving forces in the automotive industry, which was still in its early days. Benz & Cie. fitted its cars with strong engines with dual ignition and four-valve technology. This put the Prinz Heinrich cars way ahead of their time from a technical standpoint – even if they did not win the competition they were originally built for.

It is nothing short of a miracle that two of these motor cars have survived more than one hundred years since the race. The Prinz-Heinrich Benz with the race number "38" still even has the modified 5.4-litre engine fitted for the Tsar Nicholas Tour that took place in the same year.

Having been extensively and faithfully restored by experts at Mercedes-Benz Classic in collaboration with specialists from all over the world, the dark green racing car now attests to the company's pioneering and innovative spirit at major classic events around the world. It is only natural that this classic car should also take part in the one event that is for many the most important of the year: Pebble Beach.

Pebble Beach is one of the most prestigious competitions for historic vehicles worldwide, and the Concours d'Élégance marks the highlight of a weekend entirely dedicated to elegant, refined, and outstanding vehicles. Mercedes-Benz looks back on a long successful history at this sophisticated event on the Pacific coast of the USA that traditionally takes place on the third weekend in August: Since 1950, the brand's long list of awards has included several overall victories as well as more than 120 class victories and special prizes. The previous year, a Mercedes-Benz Saoutchik S Type picked up the coveted "Best of Show" title.

At the 2013 race, Mercedes presented a completely different kind of car. The fog over the Californian coastline, as much a part of Pebble Beach as the jury's panama hats, enveloped the countless gems at the 18th hole in a soft light, subduing the overwhelming designs of the car bodies, with their generous use of paint and chrome, alongside which the Prinz Heinrich – with its formal simplicity – seems almost inconspicuous. This is partly due to its fresh paintwork, which blends with the perfectly cut lawn to create a gentle shade of green that only connoisseurs can decipher – and yet it is thanks to this fresh lick of paint that the restored racing green giant is the unofficial highlight of the most important classic event of the year. •

Sportliche Form aus vergangenen Tagen. Der lange Prinz-Heinrich Wagen der einstigen Benz & Cie erinnert an einen Torpedo.

Sporty design from the past. The long, elegant shape of the Prinz Heinrich car made by Benz & Cie. is reminiscent of a torpedo.

Vier Sitze und dickes Leder sorgten für einen guten Komfort - beim Geradeauslauf. Zum Schalten musste der Fahrer nach Außen greifen.

Four seats and thick leather ensured a comfortable ride on straights. The driver had to reach outside the car to change gears.

[redacted]

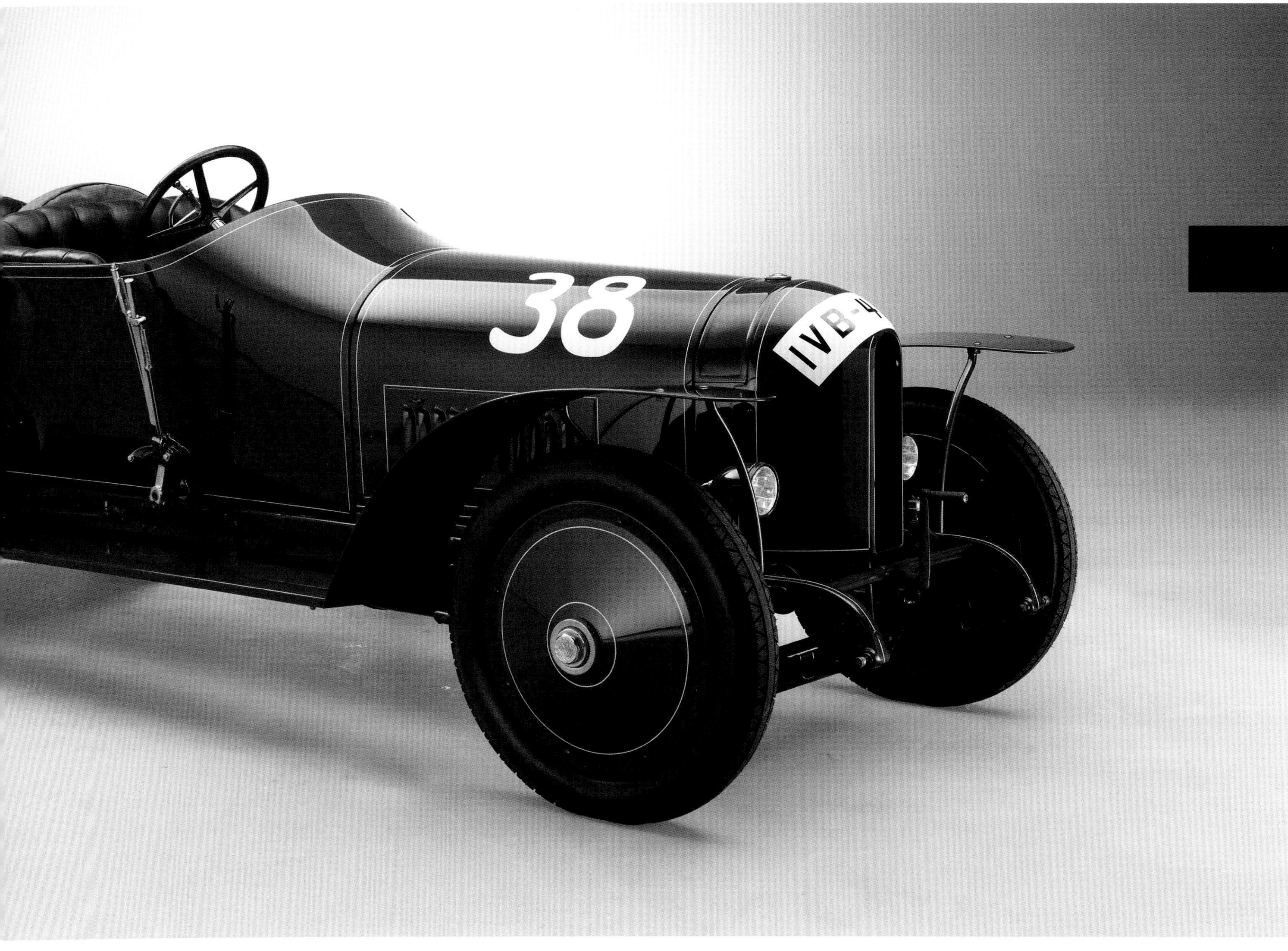
38
IVB-

1
Herkomer-Konkurrenz. Als Zuverlässigkeitsfahrt von 1905-1907 die Vorgängerveranstaltung der Prinz-Heinrich-Fahrt. Im Bild der spätere Sieger Edgar Ladenburg auf seinem Mercedes 40 PS Tourenwagen.
The Herkomer-Konkurrenz was an endurance race staged from 1905-1907 that preceded the Prince Heinrich Tour. The picture shows overall winner Edgar Ladenburg in his Mercedes 40 hp touring car.

2
Prinz-Heinrich-Fahrt 1908: Fritz Erle im Bergrennen bei Bacharach.
Prince Heinrich Tour 1908:
Fritz Erle on the hill climb near Bacharach.

3
Prinz-Heinrich-Fahrt 1908: Der spätere Gesamtsieger Fritz Erle auf seinem Benz 50 PS Tourenwagen.
Prince Heinrich Tour 1908: Overall winner Fritz Erle in his Benz 50 hp touring car.

4
Prinz-Heinrich-Fahrt 1908: Carl Neumaier auf seinem Benz 50 PS Tourenwagen.
Prince Heinrich Tour 1908: Carl Neumaier in his Benz 50 hp touring car.

5
Prinz-Heinrich-Fahrt 1908: Willy Pöge erzielte auf dem Mercedes Prinz-Heinrich-Tourenwagen den zweiten Platz in der Gesamtwertung.
Prince Heinrich Tour 1908:
Willy Pöge came second overall in his Mercedes Prinz Heinrich touring car.

6
Prinz Heinrich im Gespräch mit Fritz Erle bei der Prinz-Heinrich-Fahrt 1909.
Prince Heinrich talking to Fritz Erle at the Prince Heinrich Tour in 1909.

7
Prinz-Heinrich-Fahrt 1909: Fritz Erle auf dem Benz Tourenwagen beim Start zum Forstenrieder Park-Rennen.
Prince Heinrich Tour 1909: Fritz Erle with his Benz touring car at the start of the Forstenried Park race.

8
Fritz Erle auf einem Benz Tourenwagen bei der Prinz-Heinrich-Fahrt 1909.
Fritz Erle in a Benz touring car at the Prince Heinrich Tour in 1909.

9

Max Wild am Steuer eines Mercedes Tourenwagens bei der Prinz-Heinrich-Fahrt 1909.
Max Wild at the wheel of a Mercedes touring car at the Prince Heinrich Tour in 1909.

10

Prinz Heinrich, Stifter und Namensgeber, auf seinem Benz Tourenwagen bei der Prinz-Heinrich-Fahrt 1910.
Prince Heinrich, who sponsored the race and lent his name to it, in his Benz touring car at the Prince Heinrich Tour in 1910.

11

Einer der bei der Prinz-Heinrich-Fahrt 1910 eingesetzten Benz Spezial-Tourenwagen.
A Benz touring car made especially for the Prince Heinrich Tour in 1910.

12

Fritz Erle auf seinem Benz Spezial-Tourenwagen, mit dem er bei der Prinz-Heinrich-Fahrt 1910 den fünften Platz belegte.
Fritz Erle with his Benz special touring car which came fifth at the Prince Heinrich Tour in 1910.

13

Otto Philipp auf seinem Benz Spezial-Tourenwagen bei der Prinz-Heinrich-Fahrt 1910.
Otto Philipp in his Benz special touring car at the Prince Heinrich Tour in 1910.

14

Der von Alfred Vischer gesteuerte Mercedes Spezial-Tourenwagen in voller Fahrt bei der Prinz-Heinrich-Fahrt 1910.
The Mercedes special touring car driven by Alfred Vischer at full throttle at the Prince Heinrich Tour in 1910.

15

Der belgische Mercedes-Vertreter und Rennfahrer Théodore Pilette auf seinem Mercedes Spezial-Tourenwagen bei der Prinz-Heinrich-Fahrt 1910.
The Belgian Mercedes representative and racing driver Théodore Pilette with his Mercedes touring car at the Prince Heinrich Tour in 1910.

[redacted]

16

Mercedes-Werksfahrer Alfred Vischer am Steuer des Mercedes Spezial-Tourenwagens bei der Prinz-Heinrich-Fahrt 1910.
Mercedes company driver Alfred Vischer at the wheel of a Mercedes touring car at the Prince Heinrich Tour in 1910.

10

11

12

13

14

15

16

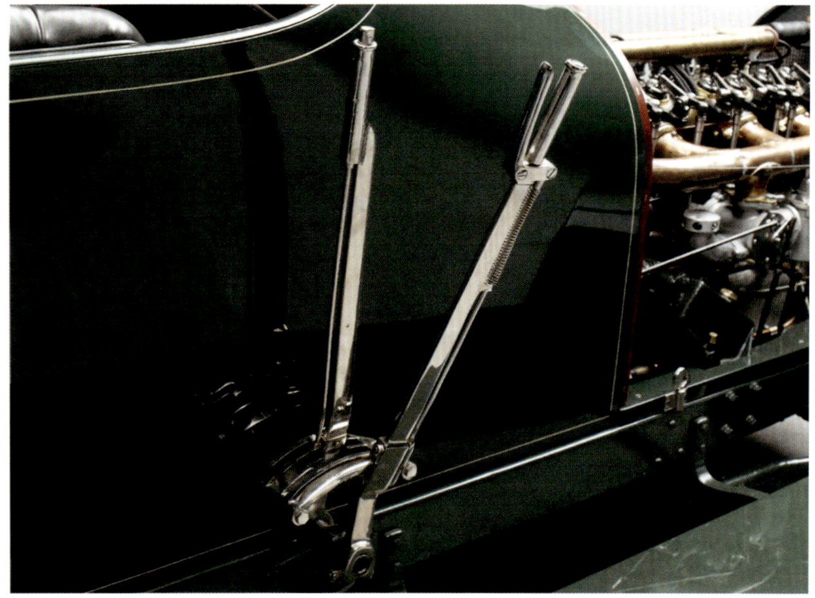

38

BÉLA

PRAGMATIKER,
QUERDENKER,
VISIONÄR.

BARÉNYI

MAVERICK, PRAGMATIST, VISIONARY.

Barényis Ideenschmiede war eine „Holzhütte" auf dem Werksgelände in Sindelfingen (s. Pfeil).
Barényi's „hothouse" was actually a wooden shack on the Daimler-Benz site in Sindelfingen (see arrow)

„Ein Haus wie Daimler-Benz kann nicht von der Hand in den Mund leben. Herr Barényi, Sie denken 15 bis 20 Jahre voraus. Sie kommen in Sindelfingen unter einen Glassturz. Was Sie erfinden, kommt direkt in die Patentabteilung."

Souverän. Smart. Konkret. Gerüstet mit scharfem Geist und unermüdlichem Idealismus war der junge Ingenieur angriffslustig in das Bewerbungsgespräch eingestiegen, hatte mit seinen konkreten Vorstellungen und visionären Ideen auf ganzer Linie überzeugt.
Dr. Wilhelm Haspel, 1938 als stellvertretendes Mitglied im Vorstand der Daimler-Benz AG, hatte die Qualitäten des damals arbeitssuchenden Ingenieurs sofort erkannt und schenkte ihm schon bei diesem Gespräch den Vertrauensbonus, der notwendig war, um später Großes leisten zu können. Eine Traumsituation für jeden Ingenieur, die angesichts der angedeuteten beruflichen Freiheiten jedem Controller erst einmal Tränen in die Augen getrieben haben dürfte.

Analytischer Scharfsinn, hohe Denkerstirn, ein Gesichtsausdruck, der den Eindruck vermittelt, dass hier jemand in Gedanken versunken ist. Barényi wirkt auf den ersten Blick exakt so, wie man sich klassischerweise einen Erfinder und Konstrukteur vorstellt. Doch er war mehr. Weit mehr. Ein globaler Vordenker, ein Pionier der Automobilgeschichte. Kein anderer beschäftige sich bis dahin grundlegend so intensiv mit der Sicherheit von Automobilen, kein anderer hat so viel zur passiven Sicherheit für den gesamten Automobilbau beigetragen wie er.

Wäre „James Bond" Testfahrer bei Mercedes-Benz, „007" die Bezeichnung für die Entwicklungsabteilung, Barényi wäre für Jahrzehnte „Q" gewesen. Denn seine Erfindungen waren wegweisend, bahnbrechend, manche schlichtweg genial. Er selber blieb aber meist bescheiden im Hintergrund und war der breiten Öffentlichkeit so gut wie unbekannt. Seine Errungenschaften hingegen nicht. Sie prägen die Automobilentwicklung bis heute. Nachhaltig.

Als Barényi bei der Daimler-Benz AG begann, dürfte er anfangs allerdings etwas über den ihm versprochenen Glassturz überrascht gewesen sein, denn der entpuppte sich als Holzbaracke auf dem Werksgelände in Sindelfingen. Dafür recht großzügig dimensioniert. So entstand dort, neben den Hauptfabrikgebäuden, die wichtigste Ideenschmiede der damaligen Daimler-Benz AG, in der Barényi seine Ideen entwickeln und an Konzepten tüfteln konnte. Meist gut gelaunt und Opernmelodien pfeifend, das inspirierte ihn. Im Laufe seines Lebens brachte er an die 2500 Patente auf den Weg. Viele davon entstanden in seiner „Schmiede".

Bekannt wurde Barényi in Fachkreisen vor allem durch sein Bestreben, das Automobil neu zu denken, es grundlegend stabiler und sicherer zu machen. Bei seinen Erfindungen setzte er seinen Schwerpunkt vor allem auf das Entwickeln von Karosserie und Fahrgestell. Zudem besaß er auch ein umfassendes Sicherheitsdenken, was für die damalige Zeit eher unüblich war. So ließ Barényi die Stabilität seiner Konstruktionen auch durch kontrollierte Crashs testen und war somit der „Ur-Vater" des modernen Crash-Tests. Bald feierte ihn die Fachwelt als Sicherheitspapst. Seine bei Mercedes-Benz in Serie gegangenen Erfindungen bilden bis heute die Grundpfeiler der passiven Sicherheit im Automobilbau.

"A company like Daimler-Benz can't live from hand to mouth. Mr. Barényi, you're thinking 15-20 years ahead. In Sindelfingen, we'll create a hothouse for you and prioritize your work: what you invent will go straight to the patent department."

Confident. Smart. Practical. Equipped with a sharp mind and untiring idealism, the young engineer threw himself enthusiastically into the job interview at Daimler-Benz, and his visionary yet realistic ideas won him a position. Dr. Wilhelm Haspel, an acting member of the Daimler-Benz Executive Board in 1938, immediately recognized the young engineer's talents and placed his trust in him – the trust Barényi needed to be able to deliver great work. It was a dream situation for any engineer: a job that promised so much professional freedom that the mere thought of it would likely bring tears to the eyes of any management accountant – but not for long, as Barényi's inventions quickly proved to be profitable...

Analytically astute, with a facial expression that conveyed the impression he was lost in thought (not least because of his furrowed brow), Barényi looked like the archetypal inventor or design engineer at first glance. But he was more. Much more: a pioneer in world automotive history. No-one else had thought so much or so comprehensively about automobile safety; no-one had done as much to improve passive safety in automotive engineering.

If James Bond had been a Mercedes-Benz test driver and if 007 had been the designation for the development department, then Barényi would have been Q. It was a role he played for decades: his inventions were seminal, groundbreaking, sometimes sheer genius. For the most part, Barényi himself – a modest man – remained in the background and was as good as unknown to the public. Unlike his achievements, which still shape automotive development today and will continue to do so.

When Barényi arrived for his first day of work at the Daimler-Benz site in Sindelfingen, Germany, he might have been surprised to discover the "hothouse" he'd been promised was actually a wooden shack, albeit a very large one. In this shack, next to the main factory buildings, Daimler-Benz's most important think tank was born – a place where Barényi could develop his ideas and refine his concepts, typically with a smile on his face or while whistling opera melodies, something that kept him inspired. Over the course of his professional life, he patented an incredible 2500 inventions, many of which came into being here in his "idea lab".

Barényi was known best in industry circles for his quest to rethink the car and make it safer and more stable on a fundamental level. His inventions focused on the body and chassis, and he had a very comprehensive approach to safety, which was unusual for the time: he even tested the stability of his designs in controlled crashes, thereby becoming the father of the modern crash test. The automotive industry soon began to celebrate him as a safety guru. Today, his inventions, series-produced by Mercedes-Benz, form the basis of passive safety in automotive engineering.

Béla Barényi took an interest in car chassis from an early age. After graduating in mechanical engineering from the Vienna Technikum, a private college, in 1926, he completed a series of precise, technical freehand drawings

Im Laufe seiner Karriere brachte Barényi an die 2.500 Patente auf den Weg.
Over the course of his career, Barényi patented around 2500 inventions

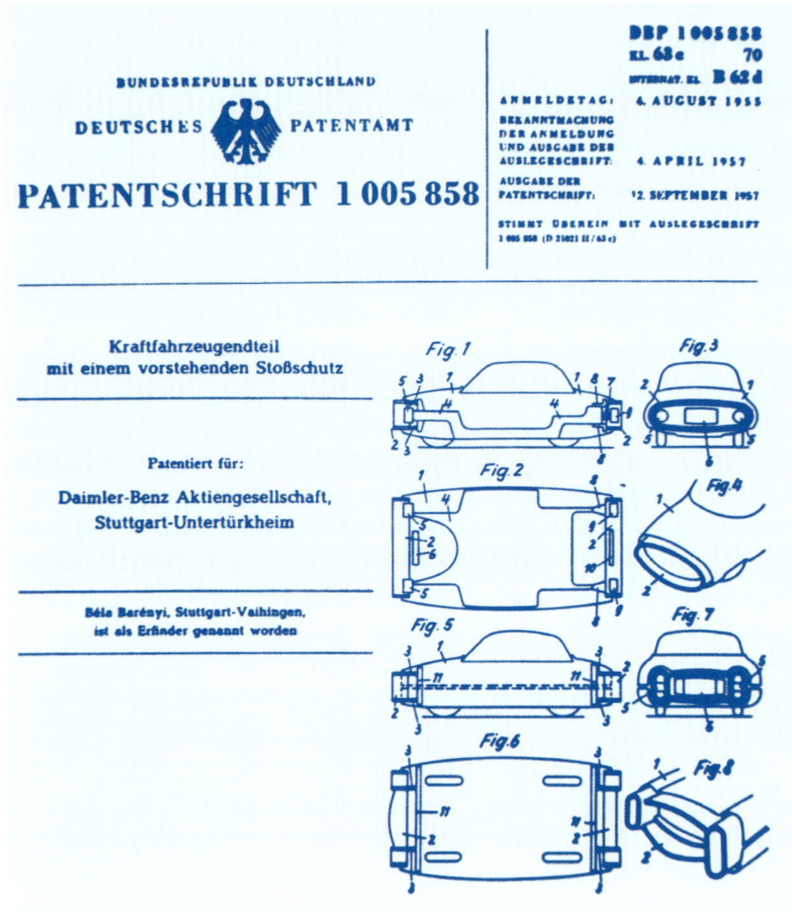

Béla Barényi begann schon in jungen Jahren, sich mit dem Automobil-Chassis zu beschäftigen. Als er 1926 das Studium des Maschinenbaus am privaten Wiener Technikum abgeschlossen hatte, entwarf er Ende der 1920er Jahre konkrete technische Freihandzeichnungen von Einzelträger-Chassis. Das Freihandzeichnen war immer schon eines seiner wichtigsten Instrumente für kreatives Schaffen. Anhand seiner Skizzen diskutierte er dann meist gleich Vor- und Nachteile.

Nachdem Barényi 1939 zur Daimler-Benz AG stieß, entwickelte er einen neuartigen Plattformrahmen mit Seitenschutz. Das war eines seiner ersten großen Projekte. Eigentlich für das Mercedes-Benz 170 V Cabriolet (W 136) vorgesehen, erreichte es jedoch erst Jahre später, 1953, im berühmten „Ponton"-Mercedes (W 120) Serienreife.

Zwischen 1945 und 1946 entwickelte Barényi Terracruiser und Concadoro. Was klingt, wie zwei Tex-Mex-Musikerhelden, sind zwei geniale Chassis-Konzepte. Mit ihnen setzte der Ingenieur seine Ideen zur passiven Sicherheit in konkreten Entwürfen um. Beim dreisitzigen Concadoro etwa dominiert die dreiteilige Bauweise mit schwenkbarer Kanzel über der Sitzreihe, der Fahrer ist hier mittig positioniert. Der sechssitzige Terracruiser besitzt hingegen eine starre Fahrgastzelle in der Mitte, die elastisch mit einer verformbaren Crashzelle an Front und Heck verbunden ist. Basierend auf dem Terracruiser, wollte Barényi 1955 das perfekte Alltagsauto entwickeln. Die Vorgaben: klein, sicher, kostengünstig. Das Projekt K-55, das intern auch unter der Bezeichnung als Vorentwicklung Pkw Version 1 oder kurz VEP 1 lief, entpuppte sich als äußerst interessantes Konstruktionsexperiment, man könnte auch sagen: Spielwiese. Mit ihm verließ Barényi erstmals seine zeichnerisch-konstruktive Basis und tauchte selbst in die spannende Welt der experimentellen Praxis ein. Sein erstes selbst gebautes Automobil besaß neben einer festen Mittelkonstruktion eine ausbaubare Antriebseinheit am Heck und einen Wagenbug, den man wechseln konnte. Die Idee war gut, der K-55 wurde aber nie realisiert, obwohl in ihm nicht nur viel von Barényis Geist, sondern auch viel Herzblut steckten. Ein Grund, warum er die Arbeiten am K-55 1973 erneut aufnahm und den K-55 verfeinerte. Aus dem Projekt entwickelte sich 1994 die Vision A, Vorläufer der Mercedes-Benz A-Klasse von 1997.

Der große Wurf im Automobilbau gelang Barényi allerdings mit einer perfektionierten Karosserie, die, mit formstabiler Fahrgastzelle und Knautschzone ausgerüstet, 1951 zum Patent angemeldet wurde. 1959 erstmals in der Oberklassen-Baureihe W 111, der „Heckflosse" realisiert. Barényis wohl berühmtestes Automobil. Die Mercedes-Benz „Heckflosse" war nicht nur Luxusliner, sondern auch das erste Autombil mit Sicherheitskarosserie. Der Clou der „Heckflosse": Bei einem Unfall absorbieren Front und Heck der Rahmenstrukturen die Crashenergie und werden verformt. Die Fahrgastzelle bleibt dabei stabil. Auch das Sicherheitslenkrad, das bei einem Unfall nicht gefährlich Richtung Fahrer schnellt, sondern starr bleibt und ihn somit schützt, hatte im W 111 Premiere.
Doch Barényi wollte alles immer besser machen, war Perfektionist. So funktionierte er auch schon gerne mal ein Serienmodell zum

Bekannt wurde Barényi in Fachkreisen für sein Bestreben, das Auto neu zu denken, es sicherer und stabiler zu machen. Auch der erste Entwurf des VW „Käfers“ stammt von Barényi.
Barényi was known in industry circles for his quest to rethink the car and make it safer and more stable. He was also responsible for the first design of the famous VW Beetle.

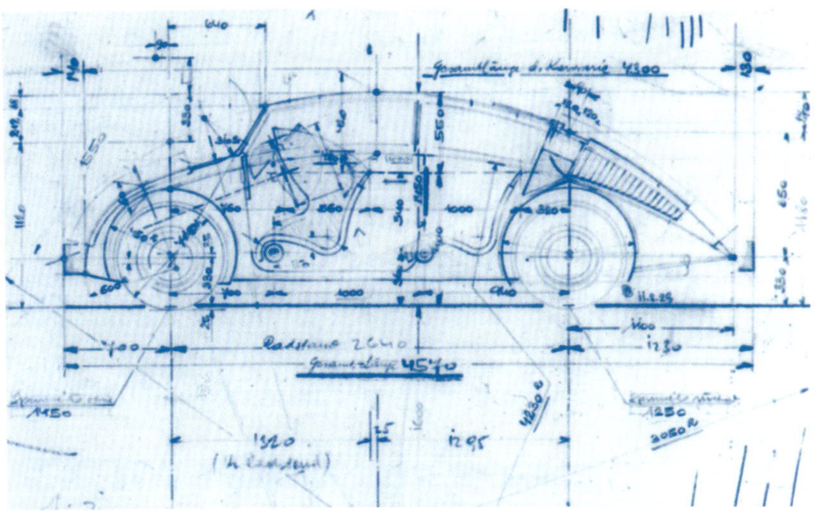

of platform chassis in the late 1920s. Freehand drawing was always one of his main creative techniques. He would sketch a design, then use the drawing as a basis for discussing its advantages and disadvantages.

After joining Daimler-Benz in 1939, Barényi developed a novel platform frame with side protection as one of his first big projects. Originally intended for the Mercedes-Benz 170 V Cabriolet (W 136), it was in fact not used until 1953, in the famous Mercedes-Benz Ponton series (W 120).

From 1945-1946, Barényi developed the Terracruiser and Concadoro. What might sound like Mexican rock bands were actually ingenious chassis concepts designed with passive safety in mind. The three-seater Concadoro boasted a three-cell design with a swiveling cockpit above the single row of seats with the driver in the middle, while the six-seater Terracruiser featured a rigid passenger cell in the center, elastically connected to a deformable crash cell at the front and rear. In 1955, Barényi wanted to develop the perfect car for everyday use based on the Terracruiser. It had to be small, safe, and good value. The K-55 project, also known internally as VEP1 (short for “Vorentwicklung Pkw Version 1” or “Car predevelopment, version 1”) was launched and turned out to be a very interesting design experiment – a true sandpit for innovation. For the first time, Barényi left his world of drawings and designs and got involved in the practical side of things. The car, the first he had built himself, boasted a solid central design with an extendable drive unit in the rear and a replaceable front. Though the idea behind it was good, the K-55 never made it into series production, despite Barényi having invested so much time and passion in it. But in 1973, he returned to the K-55 and refined it further, and later still in 1994, it ultimately evolved into Vision A, the predecessor of the Mercedes-Benz A-Class introduced in 1997.

Barényi was responsible for a major step forward in automotive engineering in 1951, when he perfected and patented a chassis divided into a crumple zone and a non-deformable passenger cell. The first series-produced car to feature this chassis was the “Fintail” – the upmarket W 111 model series launched in 1959, probably Barényi’s best-known car. More than just a luxury vehicle, the W 111 was the first automobile with a body expressly designed for safety. What made the Fintail so ingenious was that in the event of an accident, the front and rear of the vehicle absorbed the energy of the crash and were deformed, while the passenger cell remained stable. The safety steering wheel, which stayed in place instead of shooting dangerously towards the driver in the event of a crash, also made its debut in the W 111.
As a perfectionist, Barényi always wanted to make everything better. Sometimes, he would convert a series-produced model back into a “partial prototype” to test out individual parts. The W 111 was his baby, and he bestowed it with an extremely safe and stable roof design. Its convex shape made it especially stable and suitable for transporting loads: when a roof attachment was fitted, the W 111’s roof could hold ten men standing. As this description implies, Barényi tested the load-bearing capacity of the modified Fintail’s roof in an amusingly straightforward manner. To test the 220 S (W 180), he fitted a roof attachment to the

Barényi, 1948, vor einem Modell seiner sechssitzigen automobilen Zukunftsstudie „Terracruiser“.
Barényi in front of a model of his six-seater „Terracruiser“ experimental automobile in 1948

„Teil-Prototyp“ um, um einzelne Bestandteile zu testen. Wie sein Baby, den W 111. Ihm verpasste er eine äußerst stabile und sichere Dachkonstruktion, die sogenannte Pagodenkrone. Mit seiner konvexen Form sollte er besonders stabil sein und sich für einen Lastentransport eignen. Tat es auch. Barényi testete die Tragfähigkeit äußerst praktisch und unkompliziert. In Kombination mit einem Dachaufsatz hielt das Dach der modifizierten „Heckflosse“ zehn stehende Männer aus, beim 220 S (W 180) die Familie Barényi samt Freunden auf Stühlen sitzend am Tisch Kaffee trinkend. Das Konzept wurde weiterentwickelt und nahm eine konkave Form an. Das Pagodendach, von Barényi konstruiert und von Paul Bracq gezeichnet, fand schließlich beim Mercedes-Benz 230 SL (W 113) von 1963, kurz „Pagode“ genannt, Verwendung. Die „Pagode“ trat die Nachfolger des 190 SL (W 121) und des 300 SL Roadster (W 198) an. Das Pagodendach war die Krönung, der die ohnehin schon stabile Fahrzeugkonstruktion mit Fahrgastzelle und Knautschzonen noch einmal verstärkte.

Gemeinsam mit Mercedes-Benz Entwicklungsvorstand Hans Scherenberg definierte Béla Barényi 1966 die Aufteilung in aktive und passive Sicherheit, die bis heute für die ganze Automobilindustrie gilt. Konkret heißt das in eine aktive Sicherheit, die Fahraspekte betrifft und eine passive Sicherheit, in der es um Insassen- und Fußgängerschutz geht. Bélas Barényis Sicherheitsdenken schloss also auch schon damals Passanten mit ein, was für die Verhältnisse seinerzeit noch nicht üblich war. So entwickelte er neben einer Sicherheitslenkung, die 1976 im Vorläufer der heutigen E-Klasse, der Baureihe 123 Premiere hatte, auch den bekannten versenkbaren Scheibenwischer, erstmals ab 1979 in der Baureihe 126 eingesetzt.

Auch wenn Barényi in der Allgemeinheit immer noch eher ein Unbekannter ist, die Fachwelt hat den „Sicherheitspapst“ der automobilen Welt verewigt: 1994 in der Automotive Hall of Fame in Dearborn/USA und seit 2007, zehn Jahre nach seinem Tod, einen Ehrenplatz in der European Automotive Hall of Fame in Genf.

Und wer heute durch Sindelfingen fährt, der trifft dort auf das berühmte Mercedes-Benz-Werk, direkt an der Béla-Barényi-Straße gelegen. Dort, wo alles begann, dort wo irgendwo der „Glassturz“ stand und dort, wo vermutlich der damalige Leiter der Karosserieentwicklung, Karl Wilfert den jungen Barényi fragte, was seiner Meinung nach bisher falsch an den Autos konstruiert sei und er von Barényi darauf nur eine schockierende, kurze und knappe Antwort bekam: *„Vieles!“*. •

top of the car, then placed a table and chairs on it and invited his family and friends to join him for coffee – which they did. Later, the concept was developed further and the roof took on a concave form. The “pagoda top”, designed by Barényi together with Paul Bracq, first appeared in the Mercedes-Benz 230 SL (W 113) introduced in 1963. Nicknamed the Pagoda, the model was the successor to the 190 SL (W 121) and 300 SL Roadster (W 198), and the characteristic roof design increased its already high stability.

Together with Mercedes-Benz’s head of development Hans Scherenberg, Béla Barényi defined the distinction between active and passive safety in 1966 – a distinction that is standard in the automotive industry today. Active safety refers to driving safety, while passive safety concerns the protection of passengers and pedestrians. That’s right: Béla Barényi’s approach to safety also included pedestrians, which was highly unusual for the time. As well as developing safety steering, first seen in 1976 in the 123 model series – the predecessor to the E-Class – he also famously invented retractable windscreen wipers. These saw their premiere in the 126 model series in 1979.

Although Barényi’s name isn’t well known by the general public today, he’s a legendary figure in the automotive industry thanks to his status as the original safety guru. In 1994, his name was added to the Automotive Hall of Fame in Dearborn, Michigan, and in 2007, ten years after his death, he was honored with a place in the European Automotive Hall of Fame in Geneva.

Drive through Sindelfingen today, and the famous Mercedes-Benz plant – on Béla-Barényi-Strasse – is impossible to miss. That’s where everything began, where Barényi’s idea lab stood, and where Karl Wilfert, then head of chassis development, likely asked the young Barényi what he thought was wrong with contemporary car designs and received a short, shocking answer: *“A lot!”*. •

DER TERRACRUISER BESITZT EINE SICHERHEITSZELLE MIT EINER FESTEN FAHRGASTZELLE UND ELASTISCHEN KNAUTSCHZONEN.

THE TERRACRUISER FEATURED A „SAFETY CELL“ DIVIDED INTO A RIGID PASSENGER CELL AND ELASTIC „CRUMPLE ZONES“.

MERCEDES-BENZ CLASSIC INDEX

DER MERCEDES-BENZ CLASSIC INDEX (MBCI) SPIEGELT SEIT 2012 DIE WERTENTWICKLUNG 29 KLASSISCHER MERCEDES-BENZ AUTOMOBILE.

SINCE 2012, THE MERCEDES-BENZ CLASSIC INDEX (MBCI) HAS MEASURED THE VALUE OF 29 CLASSIC MERCEDES-BENZ AUTOMOBILES.

Seit Anfang 2012 gibt die Historic Automobile Group International den Mercedes-Benz Classic Index (MBCI) heraus. 29 repräsentativ ausgewählte klassische Mercedes-Benz-Modelle von der Vorkriegszeit bis zu den Emerging Classics wie dem E 500 Final Edition (W 124) aus den späten Achtzigern haben seit 1980 einen durchschnittlichen Wertzuwachs von mehr als 8 Prozent. Bis Ende November 2013 konnte der MBCI von 100 auf 135.71 Punkte zulegen. Seit Jahresanfang 2013 legte er bis November um 21.55 Prozent zu. Für den Liebhaber klassischer Mercedes-Benz Fahrzeuge ein schöner Beweis, dass sich das Investment auch lohnt.

Die Historic Automobile Group International misst seit 2009 den Wert historischer Automobile anhand von realen Transaktionen auf dem Weltmarkt. Auktionsergebnisse stellen lediglich ca. 30% der gemessenen Verkäufe dar. Die Ergebnisse fließen in den HAGI-TOP-Index ein, in dem die Marktsegmente klassischer Porsche, klassischer Ferrari und aller anderen Fahrzeuge gemessen werden. Wie bei einem Aktienindex werden die 50 repräsentativ ausgewählten Fahrzeuge nach ihrer Marktkapitalisierung gewichtet.

Schwergewicht im HAGI-TOP-Index ist der Mercedes-Benz 300 SL als Flügeltürer und als Roadster, von denen mit rund 3.000 Stück ungewöhnlich viele Autos für dieses Segment hergestellt wurden. Trotzdem notieren Flügeltürer um die Million Euro. Gute Roadster nähern sich ebenfalls dieser Schwelle. Deutlich teurer sind die seltenen Aluminiummodelle, von denen nur 29 Stück gebaut wurden. Unlängst wechselte ein Exemplar für rund 4 Millionen Euro den Besitzer.

Der Mercedes-Benz-Classic-Index misst nicht nur die besonders hochwertigen Fahrzeuge. Als Marken-Index bildet er repräsentativ die Wertentwicklung historischer Mercedes-Benz an sich. Neben dem 300 SL sind auch der 190 SL oder die 230/250/280 SL von Bedeutung. Hier rechtfertigen Preise um die

Since the start of 2012, the Historic Automobile Group International (HAGI) has published the Mercedes-Benz Classic Index (MBCI), which measures the value of a representative selection of 29 classic Mercedes-Benz cars ranging from pre-WWII models to emerging classics like the E 500 Final Edition (W 124) from the late 1980s. By November 2013, less than two years after its launch, the MBCI had risen from its starting value of 100 to 135.71 points, with an increase in value of 21.55% from January to November 2013 alone. A nice confirmation for classic Mercedes-Benz lovers that their investment makes financial sense.

HAGI first began measuring the value of historic automobiles in 2009 based on real transactions on the global market. Auctions only make up around 30% of the sales HAGI takes into account when compiling its indices, the most important of which is the HAGI Top 50 index. This index is divided into three market segments – classic Porsches, classic Ferraris, and all other vehicles. Just like a stock index, the HAGI Top 50 measures the market capitalisation of 50 classic automobiles which have been selected on a representative basis.

A big-hitter in the HAGI Top 50 is the Mercedes-Benz 300 SL (Gullwing and Roadster). Around 3000 of these cars were manufactured, an unusually large number for the market segment. Despite this, Gullwings still sell for around €1 million, and Roadsters in good condition only a little less. The rare aluminium versions, of which only 29 were made, are considerably more expensive: one recently went for around €4 million.

The Mercedes-Benz Classic Index does not just list the most valuable vehicles. As a brand index, it also shows how historic Mercedes-Benz models from various eras have appreciated in value. After the 300 SL, other important model series are the 190 SL and 230/250/280 SL. When fully restored, these automobiles are worth around €100,000 each – even more in the case of the 190 SL – so the

100.000 Euro und beim 190 SL auch darüber den Aufwand einer umfangreichen Restaurierung.

Kniffliger ist die Kaufentscheidung bei einem Vorkriegswagen. Neben den S, SS und SSK-Typen, den 500 K und 540 K berücksichtigt HAGI auch die Typen 290 und 320, die je nach Karosserieform für bis zu einer halben Million den Besitzer wechseln. Die Archive von Mercedes-Benz Classic können bei diesen Fahrzeugen die detailgenaue Originalität im Rahmen einer umfangreichen Hersteller-Expertise bestätigen. Denn damals wechselte man oft das Blechkleid im Laufe eines Autolebens. Wenn es heute noch über die originale Karosserie verfügt, ist dies wertsteigernd. Am wertvollsten sind oftmals zweisitzige Cabrios, gefolgt von solchen mit Viersitzen. Coupés müssen deutlich teurer bezahlt werden als Limousinen.

Der MBCI by HAGI misst aber auch die Ponton-Fahrzeuge aus der Nachkriegszeit und berücksichtigt die wunderschönen Flachkühler-Cabrios und Coupés der Baureihe W 111/12. Stellvertretend für sie wird der Wert des 280 SE 3.5 in diesen Karosserie-Formen gemessen. Unter Beobachtung stehen ebenfalls seltene "Emerging Classics" wie der 450 SEL 6.9, der 190 E 2.3-16 (mit seinen Ausbaustufen) oder der 500 SL (R 107) von denen unlängst ein perfektes Exemplar außergewöhnliche 54.000 Euro erzielte.

Mit dem MBCI weist HAGI schlüssig nach, dass klassische Mercedes-Benz eine gute Wertanlage sind. Wer sich mit dem Gedanken trägt, in sie zu investieren, sollte jedoch über eine gehörige Portion Fachwissen verfügen. Meistens ist guter Rat von Fachleuten nicht zu teuer. Und wer nicht eine gehörige Portion Benzin im Blut hat, sollte lieber nach anderen Investitionsmöglichkeiten Ausschau halten, für die er oder sie eine echte Passion entwickeln kann. •

restoration pays for itself. Newer emerging classics, meanwhile, have seen an average increase in value of 8% since 1980.

Things are trickier when it comes to buying a pre-WWII model. As well as the S, SS and SSK, and the 500 K and 540 K, the HAGI index also includes the Mercedes-Benz 290 and 320 models, which can change hands for up to half a million euros depending on the body. The Mercedes-Benz Classic Archive can provide experts to help value these older cars by examining them in detail and confirming their authenticity. This is necessary because in the 1930s when these cars were on the road, a car's body was often replaced over the course of its life; cars with the original bodywork are worth more. The most valuable pre-WWII models are often two-seater cabrios, followed by four-seater cabrios, while coupés are worth a lot more than saloons.

The MBCI also monitors the value of post-WWII Pontons and the beautiful Fintail cabrios and coupés of the W 111/12 model series, represented in the index by the 280 SE 3.5. Rare modern classics like the 450 SEL 6.9, the 190 E 2.3-16 (and its more powerful follow-up versions) and the 500 SL (R 107) are also listed. A 500 SL in perfect condition recently sold for a remarkable €54,000.

The MBCI proves indisputably that classic Mercedes-Benz automobiles are a good investment. It goes without saying, though, that if you're considering investing in one, you need to know what you're doing when it comes to classic cars. Good advice from experts generally doesn't come too expensive. And if you're not that enthusiastic about classic cars, you'd probably be best investing in something you really are passionate about. •

IMPRINT

PHOTOCREDITS
Tim Andrew/Octane Magazine/Dennis Publishing Ltd, Daimler AG, Daimler Art Collection Stuttgart/Berlin, Getty Images (p. 62 Moviepix/Archive Photos, p. 65 Hulton Archive/Michael Brennan, p. 67 Hulton Archives/Anwar Hussein, p. 68 Hulton Archive/Evening Standard, p. 71 NBCUniversal/NBC), Mercedes-Benz Classic Archives, Mercedes Motoring/J.G. Francis/Sean Johnstun, Siggi Müller

ISBN: 978-3-943573-10-7

EDITORIAL DIRECTION
Berthold Dörrich, Jos Bendinelli Negrone

EDITORIAL MANAGEMENT & PROJECT COORDINATION
Gabrielle Berlin, Katrin Schenk

CONCEPT
Premiummedia GmbH, Reutlingen/Germany, www.kdgroup.com

CREATIVE DIRECTION
Michael Köckritz, Jos Bendinelli Negrone

MUSIC COMPILATION
ACT Music + Visio GmbH * Co. KG

PUBLISHING COUNCIL
Michael Bock (Daimler AG), Dennis Heck (Daimler AG), Dr. Gerhard Heidbrink (Daimler AG), Matthias Restle (Daimler AG), Tobias Walburg (Daimler AG)

GRAPHIC DESIGN & LITHOGRAPHY
STAN STUDIOS GmbH & Co. KG
Wanja Drews, Jule Escherhaus, Janine Graubaum, Julian Henschel, Philipp Šille

CONTRIBUTORS
Daimler Art Collection Stuttgart/Berlin, Matthias Knebel, Iris Löhrke, mb! by Mercedes-Benz, Philipp Meier, Mercedes-Benz Classic Archives

TRANSLATION & LINGUISTIC REVIEW
Claire Gordon-Kühl, Niall McCourt, Vince Reardon, Laura Russell, Judith Diane Weston

TEXT
Berthold Dörrich (4.2, 4.4), Bernd Huesmann (4.1), Carl Christian Janke (Mercedes-Benz Classic Index), Sonja Jagdfelder (1.1), Michael Köckritz (Interview 1.3), Philipp Meier (Mythos Mercedes-Benz, 1.2, 3.1, 3.2, 3.3, 3.4, 4.3, 4.5), Manfred Prescher (2.3, 3.5), Hermann Ries (2.2), Gabriele Spangenberg (1.1), Gabriel Tamez (3.3), Dirk Vincken (1.4, 2.1, 2.2)

EDEL OPTIMAL : MEDIA

Produced by optimal media GmbH, Röbel / Germany
Printed and manufactured in Germany

earBOOKS is an imprint of Edel Germany GmbH

For more informations about earBOOKS
please visit www.earbooks.net

TRACKLIST

SIDE A

Ida Sand - Brutal Truth 3:59
(Ida Sand)
Published by ACT Publishing ltd.

Lars Danielsson - Party On The Planet 4:32
(Lars Danielsson)
Published by Edition ACT Publishing

Youn Sun Nah - Lament 3:40
(Youn Sun Nah)
Published by Edition ACT Publishing

Ulf Wakenius - Seven Days Of Falling 6:39
(Esbjörn Svensson - Dan Berglund - Magnus Öström)
Published by ACT Publishing ltd.

SIDE B

Torsten Goods - Berlin P.M. 3:48
(Torsten Goods - Jan Miserre)
Published by Edition ACT Publishing / Jan Miserre

Nils Landgren Funk Unit - You Got It 4:11
(Magnum Coltrane Price)
Published by Edition ACT Publishing

Julian & Roman Wasserfuhr - Blue Desert 4:49
(Roman Wasserfuhr - Julian Wasserfuhr)
Published by Edition ACT Publishing

Wolfgang Haffner – Heartbeat 5:24
(Wolfgang Haffner)
Published by Edition ACT Publishing

ACT